Schatzschneider

Chemistry: A Practical Course

Nils Metzler-Nolte · Ulrich Schatzschneider

Bioinorganic Chemistry
A practical course

W
DE
G

Walter de Gruyter
Berlin · New York

Authors

Prof. Dr. Nils Metzler-Nolte
Dr. Ulrich Schatzschneider
Lehrstuhl für Anorganische Chemie I
– Bioanorganische Chemie –
Ruhr-Universität Bochum
Universitätsstraße 150
D-44801 Bochum, Germany
nils.metzler-nolte@rub.de
ulrich.schatzschneider@rub.de

The cover picture shows, in the background, the X-ray crystal structure of sperm whale muscle carbonmonoxy myoglobin at 1.15 Å resolution (PDB code: 1BZR) as determined by Kachalova, Popov, and Bartunik (*Science* **1999**, 284, 473–476). The artwork was generated with Yasara 8.8.12 (E. Krieger et al., *Proteins* **2002**, 47, 39–402) and POV-Ray 3.6.1. Superimposed on this is a trace of UV/Vis spectra from the Q-band region of myoglobin observed upon binding of increasing amounts of carbonmonoxide liberated from a CORM (CO releasing molecule, J. Niesel, U. Schatzschneider et al., *Chem. Commun.* **2008**, 1798–1800). A variation of this experiment is described in Chapter 3.

This work contains 46 figures und 5 tables.

ISBN 978-3-11-020954-9

Bibliographic information published by the Deutsche Nationalbibliothek

The Deutsche Nationalbibliothek lists this publication in the Deutsche Nationalbibliografie; detailed bibliographic data are available in the Internet at http://dnb.d-nb.de.

Typesetting: Da-TeX Gerd Blumenstein, Leipzig, www.da-tex.de
Printing and binding: Strauss GmbH, Mörlenbach
Cover design: Martin Zech, Bremen

Preface and Acknowledgements

The topic for this book has emerged, evolved, and grown over the years that we have taught lab courses on *bioinorganic chemistry* (and given related lectures and seminars), first at the University of Heidelberg to students of pharmacy and molecular biotechnology, then at the Ruhr-University Bochum for students of chemistry and biochemistry. After many rounds of writing and revising scripts, printing them, and handing out collections of loose papers to our students, finally the idea for a book emerged. We are grateful to Dr. Stephanie Dawson from the publisher who shared our enthusiasm and helped the project materialize.

As this book is based on our own lab classes, we are grateful to the many co-workers, project students, and of course the students in the course itself, who gave feedback on our manuscripts, helped to improve the experiments, and in the end proof-read the manuscript. There are too many students to name them all here, but we would like to acknowledge our coworkers who helped in the lab course and during the writing of this book: Dr. Srecko Kirin (who was involved in the very early stages of this course in Heidelberg and Chapter 6), Antonio Pinto (Chapters 6, 8, and especially 9), Johanna Niesel (Chapters 3 to 5), Dr. Harmel Peindy N'Dongo (Chapter 6), Heiko Becker (Chapter 3), David Köster (Chapter 8), Katja Steinke and Lotte Holobar (Chapter 2), as well as Anna Sosniak (Chapter 7, thanks also to Dr. Michelle Salmain at ENSC de Paris for helpful discussions). Antonio, Harmel, Johanna, and Lukasz Raszeja have supervised the course in Bochum, and we are grateful for their enthusiasm and support of this project. Finally, the authors wish to thank their research groups for their patience when it was "this time of the year again" and our research needed to be grouped around instrument use by the students in this course (which usually meant late in the evening and on weekends). We are also grateful to colleagues for helpful discussions, in particular Prof. Katrin Sommer (Ruhr-University Bochum). N. Metzler-Nolte gratefully acknowledges an invitation to the ENSC de Paris as a visiting professor, during which time a major part of this book was written and finalized.

Finally, as usual when a major deadline approaches, work at nights and weekends (and during family holidays ...) becomes inevitable and we are grateful to our families for their patience and support in the last few weeks.

Nils Metzler-Nolte and Ulrich Schatzschneider

Bochum, July 2009

Table of Contents

1 Introduction

Metal ions and their complexes are essential components in all living systems. They serve to enhance the stability of secondary, tertiary, and quarternary structures of biomolecules. They are essential components in the active site of many enzymes. They are vital for signal transduction and they may serve as cofactors for enzymes. Moreover, metal ions and their complexes find use as drugs and in molecular diagnostics. It is this cornucopia of functions, and the diversity of different aspects one might study, that makes the field of *bioinorganic chemistry* so interesting and challenging. Bioinorganic chemistry is also a highly interdisciplinary field of study. Beyond chemists and biochemists, biologists will look at different angles of the field than physicists, who have yet a different perspective from people in medicinal research. This makes the field also appealing to students, who appreciate the broad range of topics, the interdisciplinary contacts, and last but not least the unusually broad range of techniques, which they perceive as a bonus in their education.

Bioinorganic chemistry as a research topic is now firmly established in many places. Beyond the single investigator, there are today joint research efforts and even centers for "Metals in Biology" and the like. There are several regular international meetings devoted to the topic and its various sub-aspects, and at least two journals are entirely devoted to the topic, the *Journal of Biological Inorganic Chemistry* and *Journal of Inorganic Biochemistry*. All indications point to a well-established field of research.

Courses in bioinorganic chemistry have become part of the university curriculum, nowadays, at least for the training of chemists and bio-chemists. Courses, lectures, and seminars on bioinorganic chemistry are usually held at the advanced Bachelor or Master stage. There are graduate schools where bioinorganic chemistry is an integral part of the curriculum. In terms of practical training, however, there seems to be much less coordinated activity. Students at the Master level are frequently "embedded" in research groups, which means that they become familiar with the techniques and systems of interest in this particular research group. What is lacking, then, is an overview of general topics and common techniques in the field. Concurrently (and maybe not entirely unrelated), there is a tendency towards structured programs (such as PhD schools) and integrated courses. It is the purpose of those activities to counter the steadily increasing individualization and specialization of research at ever earlier stages in a student's education by providing exposure to a broader background and more diverse topics than a single research group can provide.

With these observations in mind, we have written the present book. Bioinorganic chemistry provides a perfect starting point for an interdisciplinary course in the laboratory, and working on problems of bioinorganic chemistry in the laboratory almost effortlessly gives exposure to a variety of chemical, biochemical, and analytical techniques. For the student, this book complements and extends the range of good textbooks on bioinorganic chemistry which are available but will by their very nature be unspecific about details of experimental techniques or procedures. By having used these techniques in the laboratory themselves, the students will be more comfortable in research seminars or in lectures at conferences. Second, this book is intended to serve as a quick and reliable starting point for establishing a lab course. Naturally, it provides no more than a starting point that will necessarily be modified according to the requirements of the participating scientists and students.

Each chapter in this book is organized in a similar way. A brief summary of the contents is followed by learning targets in a bulleted list. The reader is then introduced to the background of the experiment, as well as guided to some specific information related to the topic of the experiment. We wish to emphasize that this book is *not* a textbook of bioinorganic chemistry, or a substitute for one. The background information is usually brief and condensed as far as general concepts are concerned. The experimentalist should be put in a position to perform the experiment competently and discuss the data that are obtained with confidence. Next, the experiment itself is described, including experimental details and an extensive list of materials (chemicals and laboratory equipment). As this book is aimed at advanced students, we assume knowledge of basic experimental procedures. Also, familiarity with safety-related issues and the use of standard procedures is assumed. Safety data for common procedures or most of the standard chemicals used are *not* provided as part of this book. Safety instructions are only given for special chemicals such as ethidium bromide or cisplatin. Their absence in this book *in no case* implies that a chemical (or procedure) is harmless! In a lab course, collecting the safety data for *all* chemicals prior to beginning the experiment could be made a compulsory part of the course work by the instructor. A number of leading references are provided at the end of each chapter. This list is certainly not comprehensive and merely a (personal) choice of references which we find useful to get students started on a given topic.

The topics covered in this book can also be used as material for a seminar on bioinorganic chemistry. The introduction for each chapter should provide enough background information and leading references to enable advanced students to work up their own seminar topic. While the choice of topics reflects the research preferences of the authors, an effort was made to present a diversity of topics, chemical synthesis techniques (organic ligands, metal complexes, special techniques), biological systems (peptides, proteins, DNA, small model compounds for enzymes) and experimental techniques (colorimetric, fluorimetric, electrochemical, chromatographic, mass spectrometric analyses, gel electrophoresis, and cell culture techniques).

Chapters 2 and 3 are related to small molecule activation and signaling (O_2, H_2O_2, CO, NO), and to small molecules as models for metallo-enzymes. Chapters 4 and 5 are centered around DNA, using metal complexes to cleave DNA or probe its structure and properties by intercalation. Chapter 6 presents the synthesis of a metal-peptide bioconjugate, and it introduces the reader to the use of organometallic systems in biology, a rapidly growing research area now called *bioorganometallic chemistry*. Chapter 7 deals with the covalent modification of proteins, at which point electron transfer in biological systems is introduced. This chapter is connected to the following Chapter 8, where electrochemical measurements on metal complexes and the conjugates from the two previous chapters are carried out, evaluated, and compared. The final Chapter 9 introduces principles of medicinal inorganic chemistry, in particular anti-tumor drugs and the testing of anti-proliferative activity on tumor cell lines.

The authors of this book strongly believe that there is a genuine need for interdisciplinary training on a practical level in many places in the world. We have therefore compiled this book to encourage the initiative and provide a quick start. On the other hand, there are as many different settings for a lab course on bioinorganic chemistry as there are institutions planning to initiate such a course. Courses will vary in length and number of participants, and also the background of the students (chemists, biochemists, (molecular) biologists, students of biotechnology or molecular medicine, and of related subjects). Most importantly, the instructor will have to define the purpose of the course, for example whether a general introduction is intended or specific topics and/or experimental techniques should be learned. Accordingly, the experiments can be adjusted in length and depth, and more (or less) theoretical background can be required. A section on variations of the experiment and additional questions has therefore been added to each experiment.

Finally, we have been asked by very senior colleagues as well as junior researchers about experimental details of experiments that we used or reported. While the primary literature is of course extensive, and individual techniques may have been described in volumes like *Methods in Enzymology*, there is no collection of "tested" experimental procedures for common experiments in bioinorganic chemistry. This book is an initial attempt towards this direction, as it collects and describes in detail a number of tested experimental procedures.

The experiments presented herein have emerged from a laboratory course on bioinorganic chemistry that the authors have taught over ten years to students with different backgrounds at different institutions. Currently, we are teaching this course to students in chemistry and biochemistry at the Ruhr-University Bochum during their Master program. Typically, six experiments out of the nine presented herein are performed by groups of two students over a period of two to three weeks. We can accommodate a maximum of six groups, but this number is mainly limited by instrument access and supervision, keeping in mind that the course runs in parallel to the research activities of the group. The presentation of the background and the experiments has evolved over the years, which is to say that they represent a tested and workable base, but they may not be optimal in every aspect. We thus welcome any corrections, additions, and suggestions for improvements to the material presented herein. Also, we will be happy to share our own experience in running a laboratory course on bioinorganic chemistry with interested colleagues and to provide detailed information on the way this course is presently organized and running in our laboratory, ideally in return for sharing your experiences.

So, whatever your intention for reading this book might be, we hope that you find the material useful and interesting, and hope that by reading through the chapters and thinking of your own experiments, you will appreciate the broad and interdisciplinary spirit of our common scientific love, bioinorganic chemistry!

Bibliography

Several textbooks and other sources of information specifically on bioinorganic chemistry are available and listed below. These will provide a much deeper background of the topics covered in this book. In addition, and depending on their experiance, readers of this book might search for more specific background information on coordination chemistry, description of spectroscopic techniques, general inorganic chemistry, or biochemistry. For all of these topics, excellent books are available from various authors and publishers.

- H.-B. Kraatz, N. Metzler-Nolte (Eds.), *Concepts and Models in Bioinorganic Chemistry*, Wiley-VCH, Weinheim 2006
- J. J. R. Fraústo da Silva, R. J. P. Williams, *The Biological Chemistry of the Elements*, 2nd ed, Oxford University Press, Oxford 2004
- I. Bertini, H. B. Gray, E. I. Stiefel, J. S. Valentine (Eds.), *Bioinorganic Chemistry*, University Science Books, Sausalito 2007
- S. J. Lippard, J. M. Berg, *Principles of Bioinorganic Chemistry*, University Science Books, Sausalito 1994
- W. Kaim, B. Schwederski, *Bioinorganic Chemistry: Inorganic Elements in the Chemistry of Life*, Wiley-VCH, Weinheim 1994
- R. M. Roat-Mallone, *Bioinorganic Chemistry: A Short Course*, 2nd ed, Wiley, New York 2007
- P. C. Wilkins, R. G. Wilkins, *Inorganic Chemistry in Biology*, Oxford University Press, Oxford 1997
- J. A. Cowan, *Inorganic Biochemistry: An Introduction*, Wiley-VCH, Weinheim 1997

The Sigel family is editing a book series on metals in biology, which has changed its title and publisher recently:

- *Metal Ions in Biological Systems* published by Marcel Dekker, New York (Vol. 1–44, Eds. Helmut and Astrid Sigel) became
- *Metal Ions in Life Sciences* published by John Wiley and Sons, Chichester (Vol. 1–4, eds. Helmut, Astrid and Roland K. O. Sigel), which recently migrated to another publisher
- *Metal Ions in Life Sciences* published by The Royal Society of Chemistry, Cambridge (from Vol. 5 on, same editors as before)

A few thematic volumes of *Chemical Reviews* are also devoted to topics related to this book:

- Vol. 96(7) "Bioinorganic Enzymology" (1996)
- Vol. 99(9) "Medicinal Inorganic Chemistry" (1999)
- Vol. 104(2) "Biomimetic Inorganic Chemistry" (2004)
- Vol. 105(6) "Inorganic and Bioinorganic Mechanisms" (2005)
- Vol. 107(10) "Hydrogen: An Overview" (2007)

Some very interesting chapters are also found in *Comprehensive Coordination Chemistry II*, Vol. 8 ("From Biology to Nanotechnology", 2004) and in *Comprehensive Organometallic Chemistry III* (2006): Vol. 1, Chapter 31 (overview of *Bioorganometallic Chemistry*), and several chapters in Vol. 12.

Two journals publish original research articles on bioinorganic chemistry exclusively:

- *Journal of Biological Inorganic Chemistry* published by Springer, Heidelberg, Germany
- *Journal of Inorganic Biochemistry* published by Elsevier, Amsterdam, The Netherlands

2 Small molecule activation and inactivation by metalloenzymes and their model compounds

Summary. In this chapter, we will discuss the role of small molecules like O_2, N_2, CO, NO, and H_2 in biochemical cycles with a particular focus on dioxygen as the basis of aerobic life. You will learn how normal and pathological processes in the respiratory chain lead to the generation of reactive oxygen species (ROS) and their dual role with benefits in inflammatory processes as well as negative impacts on cell constituents. Then, examples of metalloenzymes involved in control of cellular ROS levels will be discussed at a molecular level. In the experimental part, you will prepare a small molecule model complex of the superoxide dismutase (SOD) enzyme and study its action in ROS degradation with the aid of UV/Vis spectroscopy.

Learning targets

✓ biological role of small diatomic molecules, in particular dioxygen
✓ formation of reactive oxygen species (ROS)
✓ inactivation of ROS by the superoxide dismutase (SOD)/catalase system
✓ small-molecule models of metalloenzyme active sites

Background

Small diatomic molecules like O_2, N_2, CO, NO, and H_2 play a very important role in nature. Dioxygen and dinitrogen, for example, are the main constituents of the present day atmosphere at 21 and 78 vol.-%,[1]

---

[1] The remaining 1% is made up of noble gases (mostly argon), water vapor, carbon dioxide, and traces of dihydrogen as well as ozone.

respectively, and involved in two of the most important biogeochemical cycles. Carbon monoxide and nitric oxide are generated endogenously in the human body to act as small molecule messengers. Since this fascinating topic deserves a more detailed discussion, it is covered separately in Chapter 3. Some evolutionarily very old organisms from the time when the earth's atmosphere was reducing and devoid of dioxygen use dihydrogen for their metabolism and survive to this day, for example in anoxic lake sediments or deep sea vents. [1,2] Although the global nitrogen cycle with reduction of dinitrogen to ammonia as a key step is also a very interesting subject, involving nitrogenase as one of the most elaborate enzymatic machineries found in nature, [3,4] this chapter will focus on processes involving dioxygen.

In autotrophic organisms like plants and photosynthetic bacteria, light energy is used to split water into dioxygen, protons, and electrons. While the dioxygen is released into the atmosphere, the electrons are stored in the form of NADPH as a reducing agent and the proton gradient generated is used for production of ATP. The energy of these two species is then used, in a dark reaction, to produce carbohydrates and other complex organic molecules from carbon dioxide. Heterotrophic organisms cannot utilize carbon dioxide and thus rely on the carbohydrates produced by autotrophs as their source of carbon building blocks and energy. The oxidation of these compounds finally generates carbon dioxide together with reducing equivalents, with oxygen used as the final electron acceptor, leading to the formation of water (Fig. 2.1a). This latter process takes place in the mitochondria and electrons are transported along a gradient of electron carriers called the respiratory chain. If the reduction of dioxygen under normal and in particular under pathological conditions is not complete, semi-reduced species like superoxide (O_2^-) and peroxide (O_2^{2-}) are generated (Fig. 2.1b).

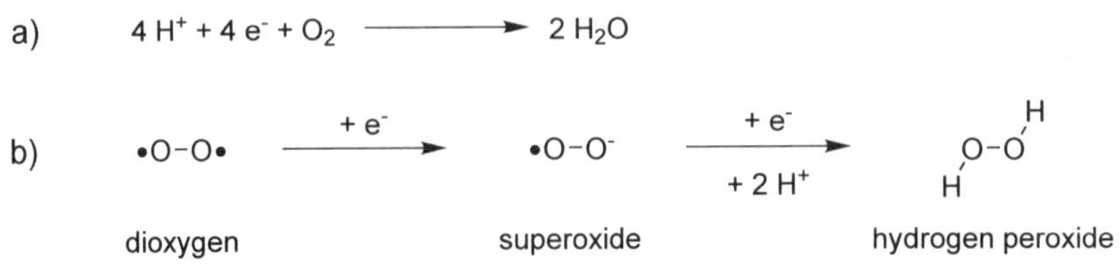

Fig. 2.1 (a) Reduction of dioxygen to water and (b) products from incomplete reduction.

In inflammatory processes they serve an important function in defense against pathogens, but under normal conditions such reactive oxygen species (ROS) leaking from the respiratory chain can damage many constituents of the cell due to redox and radical reactions. [5] Levels of ROS that are too high can lead, for example, to damage of DNA (see also Chapter 5). If this occurs beyond the repair capacity of the cell, it will lead to inheritable mutations and the development of cancer. Therefore, organisms have developed sophisticated mechanisms to control cellular ROS levels and reduce these species to less damaging compounds. In the case of superoxide, this is carried out by superoxide dismutase (SOD) enzymes. Four different classes of SODs are known which vary in their active site metal center, containing either iron (FeSOD), manganese (MnSOD), nickel (NiSOD), or copper and zinc (CuZnSOD). Since a redox-active metal is needed to carry out this reaction, the zinc in the case of the CuZnSOD is only present to modulate the properties of the copper center where the oxygen species bind. The catalytic cycle consists of two reactions. In one step, the metal center oxidizes the superoxide to dioxygen, generating a low-oxidation state active site (Fig. 2.2a). This can reduce a second equivalent of superoxide, regenerating the initial higher metal oxidation state and producing hydrogen peroxide (Fig. 2.2b).

a) $\qquad M^{n+1} + O_2^{\bullet-} \longrightarrow M^n + O_2$

b) $\qquad M^n + O_2^{\bullet-} + 2\,H^+ \longrightarrow M^{n+1} + H_2O_2$

Fig. 2.2 The two reaction steps of the general superoxide dismutase (SOD) reaction cycle. Either reaction (**a**) or (**b**) can take place first, depending on the oxidation state of the resting enzyme.

Since the H_2O_2 itself is also toxic to cells, it is further split into water and dioxygen by another class of enzymes, the catalases, which contain either heme iron groups or a dimanganese center to carry out this conversion. [6] In the CuZnSOD case, the first step is water replacement by superoxide from a $Cu^{II}(N^{His})_4(H_2O)$ center with concomitant reduction of Cu(II) to Cu(I) and release of dioxygen (Fig. 2.3a). At the same time, the imidazolate bridging the copper and zinc centers is protonated, leaving a low-coordinate copper(I) species. Then, a second equivalent of superoxide binds with reoxidation of the copper to the Cu(II) stage and protonation of the distal oxygen (the one away from the metal center) to form a copper(II) hydroperoxo complex. Addition of a second proton liberates hydrogen peroxide and closes the cycle. Thus, the CuZnSOD active site cycles

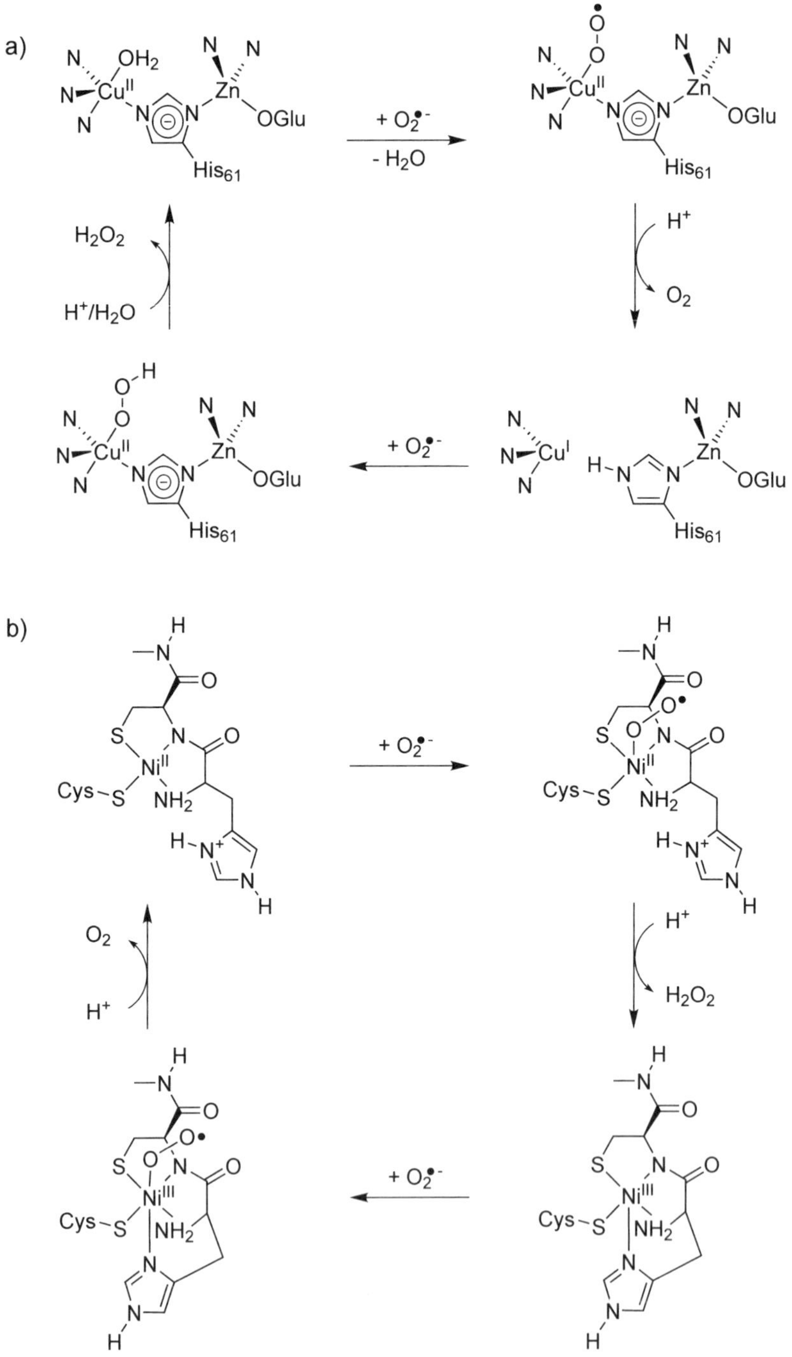

Fig. 2.3 Catalytic cycle of (**a**) CuZnSOD and (**b**) NiSOD.

between Cu(I) and Cu(II) oxidation states. In the case of the NiSOD, only a mononuclear active site is involved in which the nickel center is in a square-planar $Ni^{II}S_2N_2$ resting state (Fig. 2.3b). Upon superoxide binding, the nickel(II) is oxidized to the nickel(III) state and the addition of two protons to the reduced dioxygen species liberates hydrogen peroxide. A second equivalent of superoxide then binds to the Ni(III) center and with re-reduction to Ni(II) leads to release of dioxygen.

To control pathological ROS levels in the human body, there is significant research on artificial small coordination compounds to mimic the action of SOD enzymes. A particularly easy to prepare class of compounds are the manganese-salen complexes. [7] Condensation of two equivalents of salicylaldehyde (sal) with 1,2-ethylenediamine (en) leads to ligand H_2salen with a pre-organized square-planar N_2O_2 donor atom set (Fig. 2.4). The metal is then introduced by reaction with manganese(II)acetate in the presence of potassium chloride in air. Concomitant oxidation leads to the Mn(III)salen complex [Mn(salen)Cl], in which an additional chloride in the apical position completes the square-pyramidal coordination sphere of the metal center.

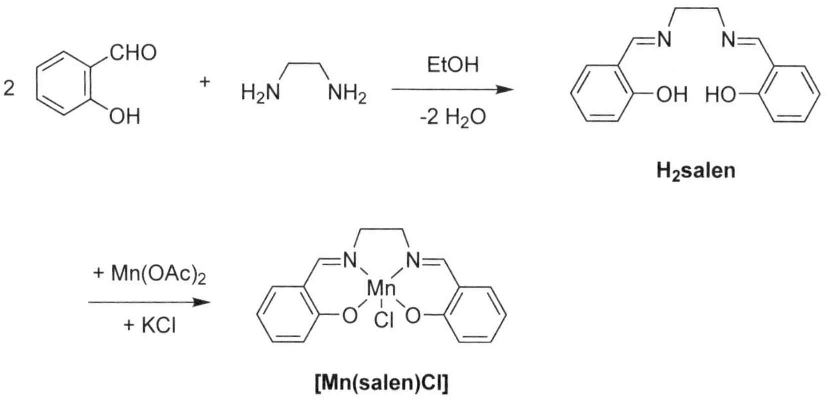

Fig. 2.4 Two-step synthesis of the manganese-salen complex [Mn(salen)Cl] used in this experiment.

In order to study the SOD activity of such a small molecule model compound, both a bio-related reaction to generate ROS and an assay system is needed. Here, we will use the conversion of hypoxanthine to xanthine by the enzyme xanthine oxidase (XO), which generates two molecules of superoxide per equivalent of hypoxanthine (Fig. 2.5). The xanthine oxidase is part of the catabolism of nucleic acids. In the first

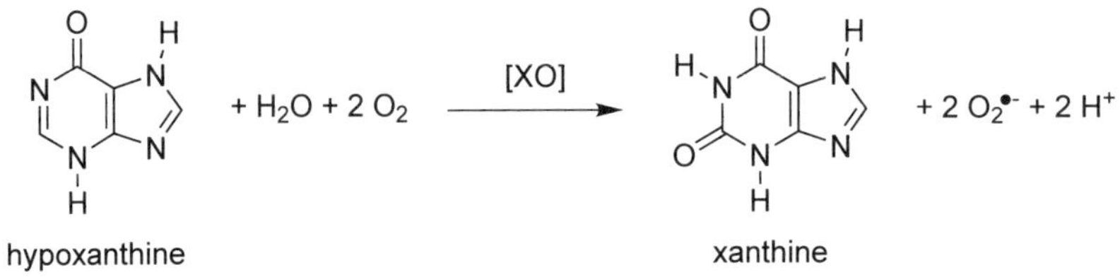

Fig. 2.5 Generation of superoxide from hypoxanthine by action of xanthine oxidase (XO).

step, adenosine monophosphate (AMP) is dephosphorylated at the $5'$-site to give adenosine. This is deaminated to form inosine which is further processed by a nucleosidase, which cleaves the N-glycosidic bond leading to liberation of the ribose sugar unit and formation of hypoxanthine. The next step is hypoxanthine oxidation by XO, which will be utilized here. The xanthine is then further oxidized to lead to uric acid, a reaction which is also catalyzed by XO.

Xanthine oxidase is also an interesting example of a metalloenzyme by itself. A dimer with a molecular weight of about 270 kDa, it harbors both a flavine (FAD) cofactor, four [2 Fe−2S] iron-sulfur centers, and a molybdenum center in the active site, which is coordinated by a special molybdopterine cofactor. During the catalytic cycle, a water-derived oxygen atom is transferred to the hypoxanthine substrate while dioxygen serves as the electron acceptor generating superoxide. [8, 9] In addition to its use as our source of ROS in this experiment, xanthine oxidase also is an important target in the therapy of gout. In this disease, defects in the purine metabolism lead to an accumulation of uric acid crystals in the joints of the skeleton and subsequent inflammatory processes. For its treatment, allopurinol is used as an inhibitor of xanthine oxidase, thus preventing the formation of insoluble uric acid.

For the detection of the generated superoxide, a UV/Vis-based assay system utilizing the nitrotetrazolium blue dye shown in Figure 2.6 will be used. In the oxidized form, this compound is yellow but reduction by superoxide leads to a blue colored species with an absorption maximum at 540 nm, which can conveniently be monitored with a spectrophotometer. [10]

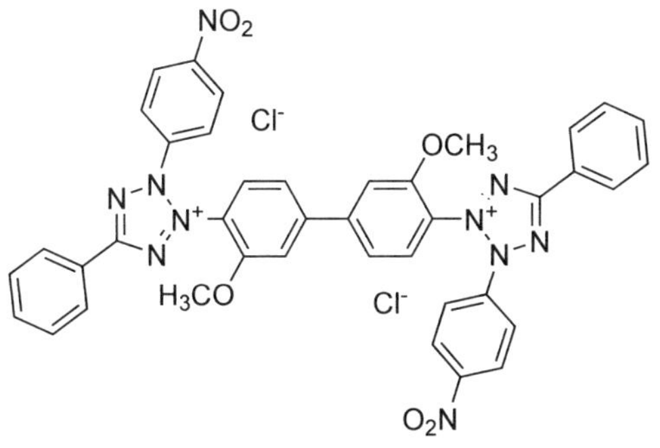

Fig. 2.6 Structure of oxidized nitrotetrazolium blue (NBT), the redox-active dye used to follow the xanthine oxidase reaction and ROS inactivation by [Mn(salen)Cl].

Experiment

Objectives

✓ sythesis of a small molecule SOD model compound
✓ enzymatic ROS generation
✓ UV/Vis spectometric determination of reaction kinetics

Materials

– salicylaldehyde ($M = 122.12\,\mathrm{g\cdot mol^{-1}}$)
– 1,2-ethylendiamine ($M = 60.10\,\mathrm{g\cdot mol^{-1}}$)
– anhydrous ethanol
– manganese(II) acetate-tetrahydrate ($M = 245.09\,\mathrm{g\cdot mol^{-1}}$)
– potassium chloride ($M = 74.55\,\mathrm{g\cdot mol^{-1}}$)
– ultrapure water
– methanol
– nitrotetrazolium blue chloride (NBT, $M = 817.64\,\mathrm{g\cdot mol^{-1}}$)
– xanthine oxidase
– hypoxanthine ($M = 136.11\,\mathrm{g\cdot mol^{-1}}$)
– disodium hydrogenphosphate ($M = 141.96\,\mathrm{g\cdot mol^{-1}}$)
– sodium dihydrogenphosphate ($M = 119.98\,\mathrm{g\cdot mol^{-1}}$)

– dimethylsulfoxide
– graded volumetric flasks (10 and 100 ml)
– pipettes and tips (10, 100, 1000 μl)
– balance ($\Delta m = 0.1$ mg or better)
– disposable single-use plastics UV/Vis cuvettes
– UV/Vis spectrophotometer
– digital stopwatch

Synthesis of bis(salicylaldehyde)ethylenediimine (H$_2$salen)

In a 500 ml three-neck flask fitted with a reflux condenser and a dropping funnel, salicylaldehyde (12.2 g, 100 mmol) is dissolved under argon in anhydrous ethanol (200 ml). Then, the dropping funnel is charged with 1,2-ethylenediamine (3.34 ml, 3.0 g, 50 mmol) dissolved in anhydrous ethanol (30 ml). At room temperature and with vigirous stirring, this solution is added dropwise to the salicylaldehyde. A bright yellow precipitate forms. If this become too difficult to stir, another 200 ml of anhydrous ethanol are added to the flask. After complete addition, stirring is continued for another 2 h. Then, the precipitated solid is filtered off on a Büchner funnel and dried in the air overnight. Yield: 10.46 g (78%). [11] ^{1}H NMR (dmso-d_6, ppm): $\delta = 13.38$ (s, 2 H, OH), 8.59 (s, 2 H, CH=N), 7.27 − 7.44 (m, 4 H, phenyl) 6.84 − 6.92 (m, 4 H, phenyl), 3.92 (s, 4 H, N−CH_2).

Synthesis of [Mn(salen)Cl]

In a 500 ml one-neck flask fitted with a reflux condenser, H$_2$salen (1.1 g, 4.1 mmol) and manganese(II)acetate-tetrahydrate (2.0 g, 8.2 mmol) are suspended in ethanol (100 ml) and heated to reflux for 3 h. Then, the solvent is completely removed in vacuo. The resulting solid is extracted with hot water (100 ml) and filtered. Solid potassium chloride (7.25 g) is added to the filtrate with stirring. The solution is stored at +4 °C overnight, leading to precipitation of a red-brown solid. The product is collected on a filter and dried in the air. Yield: 1.35 g (75%). [12] Due to its paramagnetic nature, no NMR spectrum can be obtained under standard conditions but the compound can be analyzed with IR spectroscopy and mass spectrometry: MS (FAB, 3-NBA): $m/z = 321.0\,[\text{M}-\text{Cl}]^+$, $677\,[2\,\text{M}-\text{Cl}]^+$.

Preparation of 0.1 M phosphate buffer pH 7.4

Prepare 100 ml of a 1 M solution of both disodium hydrogenphosphate and sodium dihydrogenphosphate by dissolving, in a graded volumetric flask, an appropriate amount of each salt (calculate yourself taking into account the molecular weight of both compounds) in as little ultrapure water as possible and fill up with water to 100 ml. Then mix 77.4 ml of the disodium hydrogenphosphate and 22.6 ml of the sodium dihydrogenphosphate solution in an appropriate flask.

Preparation of [Mn(salen)Cl] stock solution

Accurately weigh approximately 1 mg of [Mn(salen)Cl] on a balance, transfer to a 10 ml volumetric flask and dissolve in ultrapure water. Calculate the concentration of the stock solution using the molecular weight of the Mn-salen complex of $356.69\,\text{g}\cdot\text{mol}^{-1}$. If necessary, adjust the concentration to 0.3 mM.

Preparation of NBT stock solution

Accurately weigh approximately 20 mg of NBT chloride on a balance, transfer to a 10 ml volumetric flask and dissolve by addition of methanol (200 μl). After complete dissolution, add 0.1 M phosphate buffer pH 7.4 to 10 ml. The NBT concentration should be 2.5 to 3.0 mM.

Preparation of hypoxanthine stock solution

Accurately weigh approximately 2 mg of hypoxanthine on a balance, transfer to a 10 ml volumetric flask and dissolve by addition of dimethyl-sulfoxide (200 μl). After complete dissolution, add 0.1 M phosphate buffer pH 7.4 to 10 ml. The hypoxanthine concentration should be 1.5 mM.

Preparation of xanthine oxidase stock solution

Accurately weigh approximately 3 mg of xanthine oxidase (XO) on a balance, transfer to a glass vial and dissolve by addition of 1.0 ml of 0.1 M phosphate buffer pH 7.4. The XO concentration should be about 10 μM.

Xanthine oxidase assay

Please note: since the NBT dye tends to stick to the windows of quartz glass cuvettes and is very difficult to completely wash away, disposable single-use plastics cuvettes should be used throughout the experiment to avoid permanent contamination. Before each series of measurements, take a baseline from 400 to 700 nm on the UV/Vis spectrometer with pure 0.1 M phosphate buffer pH 7.4 according to the instruction of your instrument.

In the first experiment, you will measure the generation of ROS from hypoxanthine by xanthine oxidase. In the sample cuvette, mix 600 μl of 0.1 M phosphate buffer pH 7.4 with 200 μl of NBT stock solution (3.0 mM), 100 μl of hypoxanthine stock solution (1.5 mM), and 100 μl of xanthine oxidase stock solution (10 μM). At least a two-fold excess of NBT over hypoxanthine is needed since XO generates up to two equivalents of ROS per molecule of hypoxanthine. Record spectra in the 400 to 700 nm range in suitable time intervals until the absorption does not change anymore (accurately measure either with a stop watch or from time-stamps your instrument software automatically adds to the recorded spectra). In addition, run two control experiments in which either the xanthine oxidase or the hypoxanthine substrate are omitted from the reaction mixture. Instead, fill up with extra phosphate buffer to 1000 μl.

To study the inhibition of NBT reduction by the [Mn(salen)Cl] complex, the experiment described above is repeated but with metal complex at increasing concentration also added to the cuvette while keeping the total volume at 1000 μl. Under our conditions, the volume of [Mn(salen)Cl] stock solution added has to be varied from 0 to 50 μl to achieve a noticeable effect. Instead of scanning the whole spectra range indicated above, only the absorption at 540 nm is monitored over time in intervals of about 30 to 60 seconds.

Evaluation of results

Determine the absorption maximum of the reduced NBT dye from the end point of your measurement. Then, plot the absorption at that wavelength against the time passed since the start of the experiment. Which [Mn(salen)Cl] concentration is needed to completly suppress the ROS formation?

Variations of the experiment

With the assay described above, you can further investigate the influence of metal complex and hypoxanthine concentrations as well as temperature on the NBT reduction kinetics. The study of different substrate concentrations in the absence of a metal complex will allow you to construct a Lineweaver–Burke diagram and determine the Michaelis–Menten parameters for the xanthine oxidase. In addition, electronic influences on the SOD activity can be studied by the use of substituted salicylaldehydes with an electron-withdrawing or donating group, in particular in the 2- or 4-position to the phenolic OH group, in the synthesis of H_2salen. Also, other 1,2-diamines can be used instead of the 1,2-ethylenediamine, for example 1,2-phenylendiamine or 1,2-diaminocyclohexane. You might also wish to study other metals. For example, the manganese(II) acetate can easily be replaced by iron(II) acetate or iron(III) chloride hexahydrate to obtain the corresponding ferrous or ferric complexes. [13]

Additional questions

✓ Which other enzymes also utilize molybdopterine cofactors related to the one in xanthine oxidase (XO)?
✓ What is the role of the iron-sulfur centers in the enzyme?
✓ What is the structure of allopurinol and how does it inhibit xanthine oxidase?
✓ Why does the color of the NBT dye change from yellow to blue upon reduction?

Bibliography

[1] M. Frey, Hydrogenases: Hydrogen-activating enzymes, *ChemBioChem* **2002**, 3, 153–160.
[2] *Chem. Rev.* **2007**, 107(10), special issue on dihydrogen.

[3] J. B. Howard, D. C. Rees, Structural basis of biological nitrogen fixation, *Chem. Rev.* **1996**, 96, 2965–2982.

[4] B. M. Hoffmann, D. R. Dean, L. C. Seefeldt, Climbing nitrogenise: Toward a mechanism of enzymatic nitrogen fixation, *Acc. Chem. Res.* **2009**, 42, 609–619.

[5] H. Sies, Biochemistry of oxidative stress, *Angew. Chem. Int. Ed.* **1986**, 25, 1058–1071.

[6] A. J. Wu, J. E. Penner-Hahn, V. L. Pecoraro, Structural, spectroscopic, and reactivity models for the manganese catalases, *Chem. Rev.* **2004**, 104, 903–938.

[7] M. Baudry, S. Etienne, A. Bruce, M. Palucki, E. Jacobsen, B. Malfroy, Salen-manganese complexes are superoxide dismutase-mimics, *Biochem. Biophys. Res. Commun.* **1993**, 192, 964–968.

[8] R. Hille, The mononuclear molybdenum enzymes, *Chem. Rev.* **1996**, 96, 2757–2816.

[9] R. Hille, Molybdenum and tungsten in biology, *Trends Biochem. Sci* **2002**, 27, 360–367.

[10] M. Younes, U. Weser, Inhibition of nitroblue tetrazolium reduction by cuprein (superoxide dismutase), $Cu(tyr)_2$ and $Cu(lys)_2$, *FEBS Lett.* **1976**, 61, 209–212.

[11] S. R. Doctrow, K. Huffman, C. B. Marcus, G. Tocco, E. Malfroy, C. A. Adinolfi, H. Kruk, K. Baker, N. Lazarowych, J. Mascarenhas, B. Malfroy, Salen-manganese complexes as catalytic scavengers of hydrogen peroxide and cytoprotective agents: Structure-activity relationship studies, *J. Med. Chem.* **2002**, 45, 4549–4558.

[12] L. J. Boucher, Manganese Schiff's base complexes II – synthesis and spectroscopy of chloro-complexes of some derivatives of (salicylaldehydeethylenedi-imato)manganese(III), *J. Inorg. Nucl. Chem.* **1974**, 74, 531–536.

[13] A. Hille, I. Ott, A. Kitanovic, I. Kitanovic, H. Alborzina, E. Lederer, S. Wölfl, N. Metzler-Nolte, S. Schäfer, W. S. Sheldrick, C. Bischof, U. Schatzschneider, R. Gust, [N,N'-Bis(salicylidene)-1,2-phenylenediamine]metal complexes with cell death promoting properties, *J. Biol. Inorg. Chem.* **2009**, 14, 711–725.

3 Carbon monoxide and nitric oxide as small molecule messengers

Summary. In this chapter, you will be introduced to the biological role of carbon monoxide (CO) and nitric oxide (NO) as important small molecule messengers in the body. You will learn about the endogenous enzymatic generation of CO and NO in the cell and the signal transduction pathways influenced by these molecules. Then, the role of synthetic transition metal complexes as a source of carbon monoxide and nitric oxide for bioanalytical and therapeutic applications will be discussed. In the experimental part, you will learn how to prepare such metal complexes and how to study their thermal and photochemical CO and NO release properties with a heme-based UV/Vis assay system.

Learning targets

✓ CO and NO as signaling molecules in the body
✓ generation of CO and NO in cells and their biological targets
✓ synthetic metal complexes as a source of carbon monoxide and nitric oxide
✓ heme proteins as UV/Vis sensors for CO and NO

Background

To many people, carbon monoxide (CO) and nitric oxide (NO) are only known as highly toxic gases and common air pollutants. It is, however, now well-established that, at low concentrations, they also act as important small molecule messengers in many organisms, including humans. Both are endogenously produced in a carefully controlled manner by enzymatic reactions and serve important physiological functions, mostly protection of the organism against oxidative stress (see also Chapter 2).

Although historically preceded by studies on the biological properties of carbon monoxide, nitric oxide has received considerably more attention in a biological context, [1–3] highlighted by its nomination as "Molecule of the Year" in 1992 [4] and the 1998 Nobel Prize in Medicine to Furchgott, Ignarro, and Murad for their work on nitric oxide as a signaling molecule in the cardiovascular system. [5–7] The generation of NO in higher organisms is a two-step process starting from the amino acid L-arginine (Fig. 3.1). Both steps are catalyzed by the heme enyzme NO synthase (NOS) which leads to the formation of L-citrulline in addition to NO. Three different isoforms of NO synthase are expressed depending on the tissue. Neuronal NOS (nNOS) is found in neuronal cells and skeletal muscles, while the endothelial NOS (eNOS) is mostly localized in endothelial and epithelilal cells, but also some neurons. Finally, the inducible NOS (iNOS) is active in many cells like macrophages, hepatocytes, astrocytes, and smooth muscle cells. Also, nitric oxide can be generated in the organism non-enzymatically from a variety of other compounds such as organic nitrates, nitrites, and nitroso compounds.

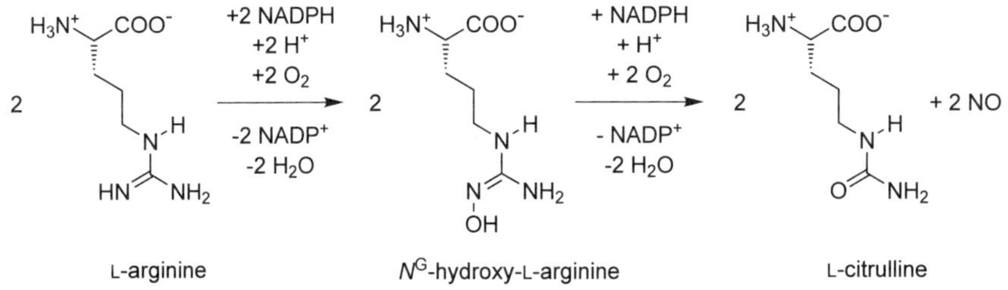

Fig. 3.1 Enzymatic generation of nitric oxide by NO synthase (NOS); the enzyme is involved in both steps.

In contrast, the biological source of carbon monoxide is not amino acid-based. Rather, most of the CO produced endogenously results from the oxidation of the porphyrin ring system in heme. [8–10] This process is mediated by the heme oxygenase (HO) family of enzymes and has been termed "the most visible of all enzyme reactions" since the highly colored products of HO action are clearly visible in the skin during the development of bruises. [11] The reaction is unique because the porphyrin group in the heme oxygenase serves as the cofactor for its own degradation. An initially formed peroxy-heme (Fig. 3.2) is thought to regioselectively attack the α-position of the ring system, leading to α-meso-hydroxyheme. After addition of a second equivalent of dioxygen, this is oxidized to verdoheme coupled with liberation of CO originating from the α-carbon. A third equivalent of dioxygen is then consumed upon further oxidation to the linear tetrapyrrol biliverdin. The whole process also consumes two equivalents of NADPH. In the final step, which involves biliverdin oxidase instead of HO, this intermediate is converted into bilirubin. Of the three heme oxygenase isoforms, HO–1 is inducible and stimulated in response to oxidative stress. In contrast, HO–2 shows some basal activity and is involved in neurotransmission and regulation of vascular tone in the brain, liver, and endothelium. The precise function of HO–3 is unknown at present.

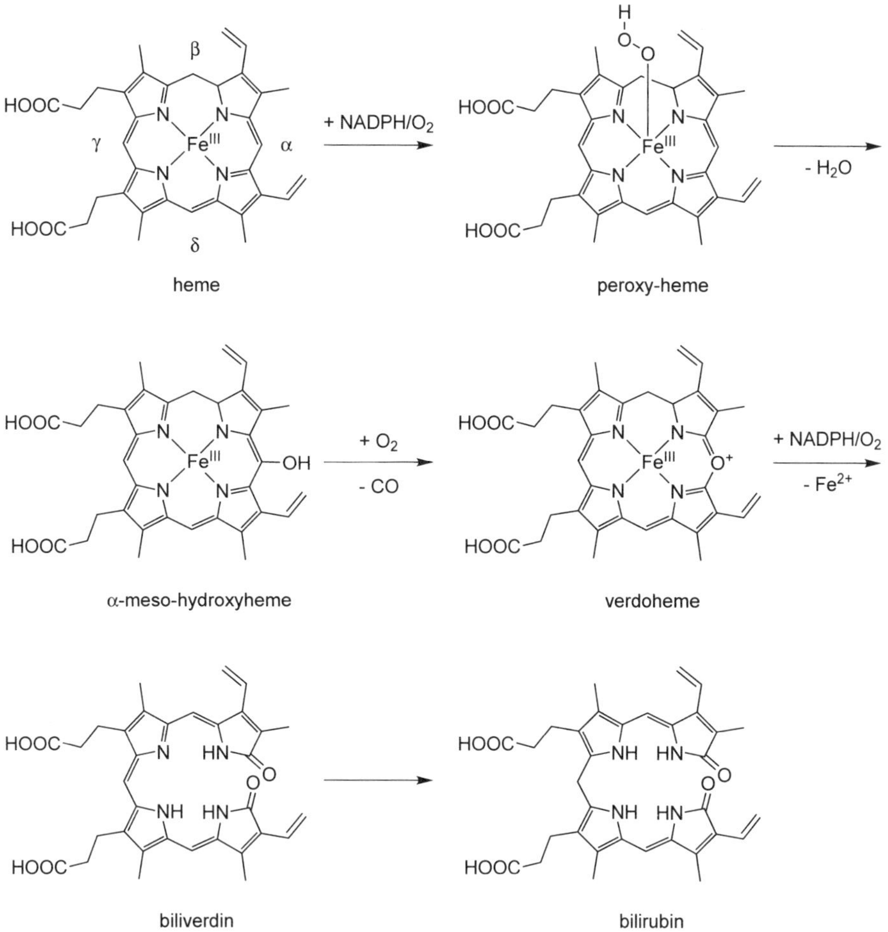

Fig. 3.2 Enzymatic generation of carbon monoxide and biliverdin by heme oxygenase (HO), which is involved in the first four steps while the final oxidation of biliverdin to bilirubin is catalyzed by a second enzyme, biliverdin oxidase.

For nitric oxide, soluble guanylyl cyclase (sGC) is thought to be the primary biological target. This enzyme catalyzes the conversion of guanosine triphosphate (GTP) to cyclic guanosine monophosphate (cGMP), which is an important second messenger and activates a number of cellular signaling pathways (Fig. 3.3). On a molecular basis, this activity is triggered by NO binding to the heme group of sGC. This leads to the displacement of a coordinated histidine which then induces major changes in the enzyme structure and enables the GTP to cGMP conversion. [12–14] In contrast, there are still considerable uncertainties regarding the precise mechanism of action of carbon monoxide. Although CO is also known to bind to sGC, the degree of activation is almost two orders of magnitude lower than that induced by NO.

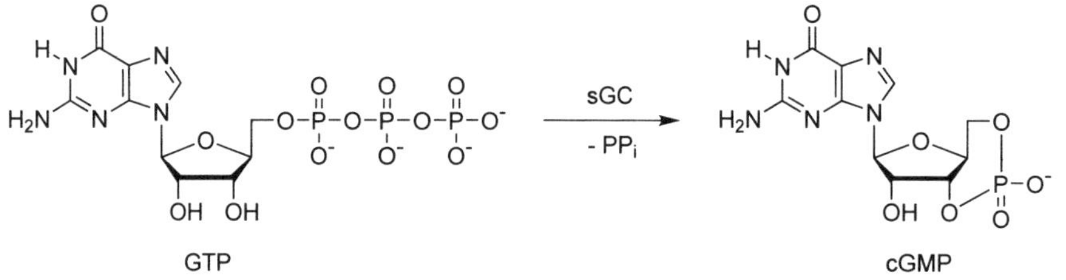

Fig. 3.3 Enzymatic conversion of guanosine triphosphate (GTP) to cyclic guanosine monophosphate (cGMP) by soluble guanylyl cyclase (sGC), an important second messenger.

Both carbon monoxide and nitric oxide are gases and as such difficult to utilize for therapeutic applications in humans, since they are highly toxic when overdosed. Thus, there is considerable interest in solid "storage forms" of CO and NO, from which the diatomic molecules can be liberated in a controlled manner. [15–18] In the case of nitric oxide, a number of organic sources for NO have been applied for over a century, ironically including the prescription of nitroglyerin to Alfred Nobel for the treatment of his heart disease. In the context of bioinorganic chemistry, metal-nitrosyl complexes have also received a great deal of attention for quite a while. For example, sodium nitroprusside ($Na_2[Fe(CN)_5(NO)]$) is used in intensive care medicine for the rapid reduction of arterial hypertension since it spontaneously releases nitric oxide at physiological pH values. For a precise spatial and temporal control of NO release and biological action, photoactivatable metal nitrosyl complexes have become the focus of considerable research in recent years. [19] In particular, iron and ruthenium complexes with different chelating ligands have been

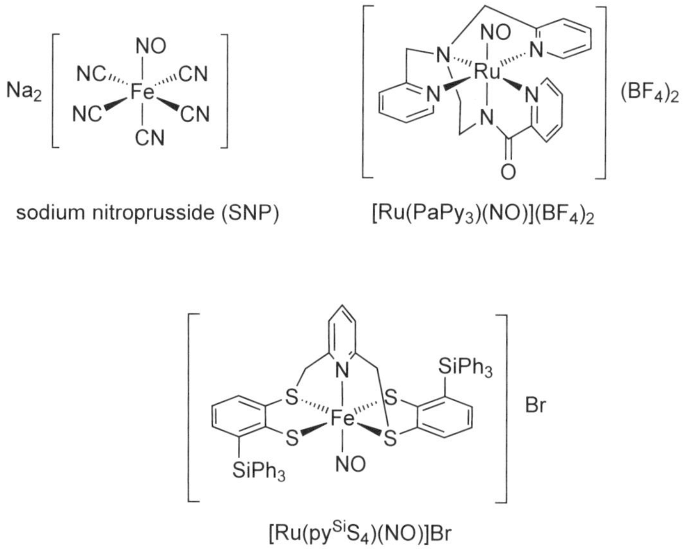

Fig. 3.4 Examples of NO releasing molecules.

prepared and studied (Fig. 3.4). In addition, there is also a steadily growing interest in CO releasing molecules (CORMs) as a stable solid storage form for carbon monoxide. Transition metal carbonyl complexes are a natural choice for the development of such CORMs and a number of manganese, iron, molybdenum, and especially ruthenium carbonyl complexes have been studied in this context (Fig. 3.5). Most of them are relatively simple compounds which allow for little structural variation, such as the most commonly employed [RuCl(glycinato)(CO)$_3$] (CORM-3). [20, 21] Only very recently, some pyrone complexes of iron and molybdenum carbonyls have been reported in which substituents on the η^1- or η^4-coordinated pyrone ring can be modified to influence the CO release properties. [22] In addition, a novel photoactivatable [Mn(CO)$_3$(tpm)]$^+$-based CORM (tpm = tris(pyrazolyl)methane) with cytotoxic activity against HT-29 human colon carcinoma cells has been identified and modified with peptides for a targeted delivery to cellular systems. [23, 24] A single main group element CO releasing molecule (CORM-A1) based on sodium boranocarbonate (Na$_2$[H$_3$BCO$_2$]) which decomposes in water to carbon monoxide and boric acid was also recently discovered. [25]

In order to study CO and NO release from metal complexes and its dependence on variables such as solvent, temperature, and pH, as well as

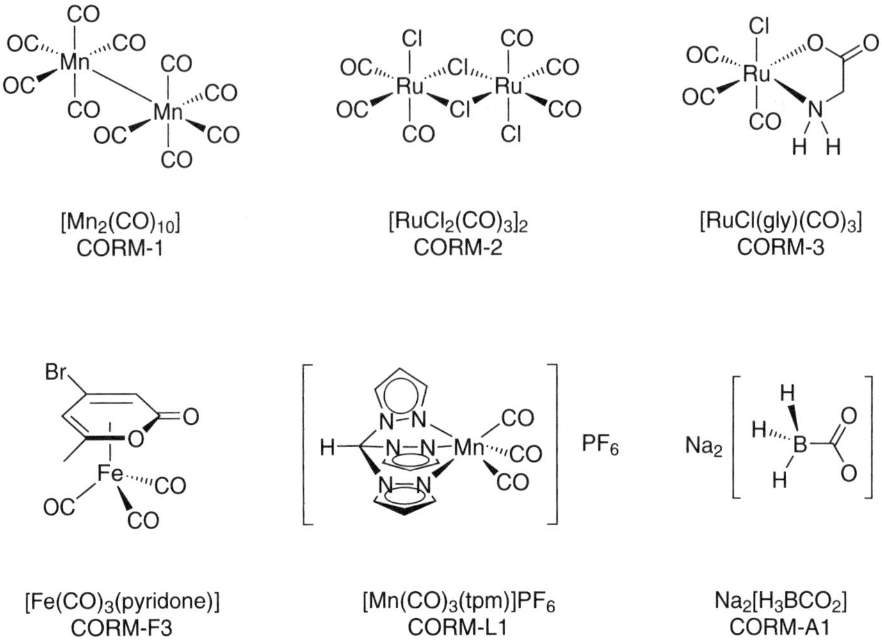

Fig. 3.5 Important CO releasing molecules (CORMs).

irradiation time and wavelength, a number of different detection methods can be employed. The most direct quantification is based on gas chromatography coupled to mass spectrometry (GC/MS), but specialized and expensive instrumentation is needed. For NO, electrochemical methods with nitric oxide-sensitive electrodes are commercially available, while amperometric CO detection is at present not mature enough for routine applications. In the experiments described in this chapter, we will instead use UV/Vis spectroscopy as the detection method. The characteristic spectral changes in the Soret- and Q-band region of heme proteins upon binding of carbon monoxide and nitric oxide can be used to follow the time course of the reaction and also to quantify the amount of CO and NO released. This system also constitutes a very simple example of a protein-based assay system and will, in addition, introduce you to some important properties of heme proteins and their small molecule adducts.

Experiment

Objectives

✓ preparation of protein stock solutions
✓ adjustment of concentrations with the aid of UV/Vis measurements
✓ photochemical reactions under an inert atmosphere
✓ determination of reaction kinetics with UV/Vis spectroscopy

Materials

– disodium hydrogenphosphate dihydrate ($M = 177.99\,\mathrm{g \cdot mol^{-1}}$)
– sodium dihydrogenphosphate ($M = 119.98\,\mathrm{g \cdot mol^{-1}}$)
– horse skeletal muscle myoglobin ($Md \sim 17\,600\,\mathrm{g \cdot mol^{-1}}$)
– disodium nitroprusside dihydrat ($M = 297.95\,\mathrm{g \cdot mol^{-1}}$)
– ultrapure water
– graded volumetric flasks (10 and 100 ml)
– pipettes and tips (10, 100, 1000 µl)
– single- or double-beam UV/Vis spectrometer
– tightly sealable UV/Vis Quartz cuvettes
– UV hand lamp $\lambda = 365\,\mathrm{nm}$ (for example as used to visualize TLC plates)
– balance ($\Delta m = 0.1\,\mathrm{mg}$ or better)
– digital stopwatch

Preparation of 0.1 M phosphate buffer pH 7.4

Prepare 100 ml of a 1 M solution of both disodium hydrogenphosphate and sodium dihydrogenphosphate by dissolving, in a graded volumetric flask, an appropriate amount of each salt (calculate yourself taking into account the molecular weight of both compounds) in as little ultrapure water as possible and fill up with water to 100 ml. Then mix 77.4 ml of the disodium hydrogenphosphate and 22.6 ml of the sodium dihydrogenphosphate solution in an appropriate flask and mix by shaking. Finally dilute 1 : 10 to obtain the 0.1 M stock solution needed.

Preparation of myoglobin stock solution

Weigh about 20 mg of horse skeletal muscle myoglobin on the balance and transfer the solid to a dry volumetric flask (10 ml). Dissolve the protein by addition of 0.1 M phosphate buffer pH 7.4. Then determine the myoglobin concentration of this solution by UV/Vis spectroscopy: measure the absorption at 408 nm ($\varepsilon_{408\,nm} = 188\,000\,l\cdot mol^{-1}\cdot cm^{-1}$), dilute the solution if necessary to ensure that $A \sim 1$, and use Lambert–Beer's law to calculate the concentration. This solution is stable for several days at room temperature if protected from light.

Preparation of sodium nitroprusside stock solution

Accurately weigh 100 mg of disodium nitroprusside dihydrate, transfer the solid to a dry volumetric flask (10 ml) and dissolve by addition of 0.1 M phosphate buffer pH 7.4 to the calibration mark. Then calculate the concentration of this solution from the molecular weight of the disodium nitroprusside dihydrate, the amount of it placed in the volumetric flask, and the volume of buffer added. This solution should be freshly prepared before the start of the experiment.

Preparation of myoglobin/sodium nitroprusside mixture and dark control

Take a baseline on the UV/Vis spectrometer using pure 0.1 M phosphate buffer pH 7.4 according to the instructions for your instrument. Then, mix stock solutions of myoglobin and sodium nitroprusside in a 1 cm Quartz cuvette and dilute with 0.1 M phosphate buffer pH 7.4 to ensure that $A_{408\,nm} \sim 1$ and $c(\text{myoglobin}) : c(\text{nitroprusside}) = 1 : 5$. Make sure the cuvette is tightly sealed, for example with a Teflon stopper. Then, place it back in the UV/Vis spectrometer, in parallel to a reference cuvette filled with 0.1 M phosphate buffer pH 7.4 if you have a double-beam instrument. If possible, keep the temperature at $+20\,°C$ and record four spectra in intervals of 15 min in the range of 300 to 700 nm, the first one directly after placing the cuvette in the instrument.

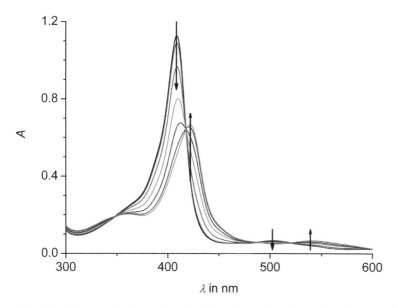

Fig. 3.6 Spectral change in the Soret-band region of a mixture of myoglobin and sodium nitroprusside upon irradiation at 365 nm.

Irradiation of sodium nitroprusside in the presence of myoglobin

Remove the cuvette from the UV/Vis spectrometer and irradiate at 365 nm with a UV hand lamp placed perpendicular to the cuvette at a distance of 10 cm while keeping the light in the room dim. The irradiation is interrupted every 10 to 15 min to record an UV/Vis spectrum in the range of 300 to 700 nm until three consecutive measurements show no further spectral changes. Use either the internal timestamps of the files store by the computer controlling your spectrophotometer or a digital stopwatch to keep the time. UV/Vis spectra from such a series of measurements are shown in Figure 3.6 as an example.

Evaluation of results

In a single diagram, plot all spectra recorded in the range of 300 to 700 nm and determine the positions of peak maxima as well as isosbestic points. Extract the absorptions at 408 and 420 nm and plot them against the irradiation time. Discuss the observed kinetics. Which reaction steps are involved and which one is rate-determining?

Variations of the experiment

Alternatively, the NO liberation from sodium nitroprusside can also be induced chemically by addition of an excess of sodium ascorbate to the mixture of sodium nitroprusside and myoglobin in the cuvette. In this case, the first UV/Vis spectrum should be measured immediately after addition of the ascorbate. If it is possible to carry out both variants of the experiment, discuss the differences in the kinetics for the chemical and photochemical NO release.

The procedure reported above can also be utilized to study other nitrosyl metal complexes or CO releasing molecules (CORMs). In the latter case, the myoglobin has to be reduced by addition of an excess of sodium dithionite prior to the experiment, as carbon monoxide only binds to the ferrous (Fe^{II}) form of myoglobin. Also, the solution in the cuvette has to be carefully degassed by bubbling with an inert gas (nitrogen or argon) prior to the measurements and kept tightly sealed to prevent re-oxidation. In studying CORMs, UV/Vis spectra should be collected in the range of 500 to 700 nm, since changes in this Q-band region are more pronounced than in the Soret-band region with its maximum around 400 nm. To study the time course of the spectral changes upon CO release, plot the absorption at 542, 557, and 577 nm against the irradiation time. If you make sure that myoglobin is present in excess, you can calculate the number of CO equivalents liberated from your compound from the plateau value of the absorption reached after exhaustive photolysis.

Some of these compounds can be easily prepared by one- or two-steps procedures. For example, CORM-3 ([RuCl(glycinato)(CO)$_3$]) is accessible in high yield from the commercially available tricarbonyldichlororuthenium(II) dimer [Ru(CO)$_3$Cl$_2$]$_2$ (CAS no. 22594-69-0) by stirring with glycine in degassed methanol in the presence of sodium ethoxide at room temperature for 18 h, followed by precipitation. [21] Photoactivatable CORM-L1 ([Mn(CO)$_3$(tpm)]PF$_6$) can be synthesized following a published procedure by reaction of tpm (tris(pyrazolyl)methane) and manganese pentacarbonylbromide in acetone. The required tpm ligand is prepared from pyrazole in good yield. [23]

Additional questions

✓ What is the structure of myoglobin, in particular its active center, and its normal function in the human body?

✓ Compare the binding and structure in nitrosyl- and carbonyl-metal complexes. Which differences can be observed?

✓ Explain the {M-NO}^{n} formalism of Enemark and Feltham for metal-nitrosyl complexes. Which structural and spectroscopic features are characteristic for the different metal oxidation states?

Bibliography

[1] J. K. S. Møller, L. H. Skibsted, Nitric oxide and myoglobins, *Chem. Rev.* **2002**, 102, 1167–1178.

[2] P. G. Wang, M. Xian, X. Tang, X. Xu, Z. Wen, T. Cai, A. J. Janczuk, Nitric oxide donors: Chemical activities and biological applications, *Chem. Rev.* **2002**, 102, 1091–1134.

[3] J. A. McCleverty, Chemistry of nitric oxide relevant to biology, *Chem. Rev.* **2004**, 104, 403–418.

[4] E. Culotta, D. E. Koshland, NO news is good news, *Science* **1992**, 258, 1862–1865.

[5] R. F. Furchgott, Endothelium-derived relaxing factor: Discovery, early studies, and identification as nitric oxide, *Angew. Chem. Int. Ed.* **1999**, 38, 1870–1880.

[6] L. J. Ignarro, Nitric Oxide: A unique endogenous signaling molecule in vascular biology, *Angew. Chem. Int. Ed.* **1999**, 38, 1882–1892.

[7] F. Murad, Discovery of some of the biological effects of nitric oxide and its role in cell signaling, *Angew. Chem. Int. Ed.* **1999**, 38, 1856–1868.

[8] P. R. Ortiz de Montellano, The mechanism of heme oxygenase, *Curr. Opin. Chem. Biol.* **2000**, 4, 221–227.

[9] L. Wu, R. Wang, Carbon monoxide: Endogenous production, physiological functions, and pharmacological applications, *Pharmacol. Rev.* **2005**, 57, 585–630.

[10] S. W. Ryter, J. Alam, A. M. K. Choi, Heme oxygenase-1/carbon monoxide: From basic science to therapeutic applications, *Physiol. Rev.* **2006**, 86, 583–650.

[11] T. R. Johnson, B. E. Mann, J. E. Clark, R. Foresti, C. J. Green, R. Motterlini, Metal carbonyls: A new class of pharmaceuticals?, *Angew. Chem. Int. Ed.* **2003**, 42, 3722–3729.

[12] S. P. L. Cary, J. A. Winger, E. R. Derbyshire, M. A. Marletta, Nitric oxide signaling: No longer simply on or off, *Trends Biochem. Sci* **2006**, 31, 231–239.

[13] S. Aono, Metal-containing sensor proteins sensing diatomic gas molecules, *Dalton Trans.* **2008**, 3137–3146.

[14] E.M. Boon, M.A. Marletta, Ligand discrimination in soluble guanylate cyclase and the H-NOX family of heme sensor proteins, *Curr. Opin. Chem. Biol.* **2005**, 9, 441–446.

[15] R. Motterlini, B.E. Mann, R. Foresti, Therapeutic applications of carbon monoxide-releasing molecules, *Expert Opin. Invest. Drugs* **2005**, 14, 1305–1318.

[16] J. Boczkowski, J.J. Poderoso, R. Motterlini, CO-metal interaction: Vital signaling from a lethal gas, *Trends Biochem. Sci* **2006**, 31, 614–621.

[17] B.E. Mann, R. Motterlini, CO and NO in medicine, *Chem. Commun.* **2007**, 4197–4208.

[18] R. Alberto, R. Motterlini, Chemistry and biological activities of CO-releasing molecules (CORMs) and transition metal complexes, *Dalton Trans.* **2007**, 1651–1660.

[19] M.J. Rose, P.K. Mascharak, *Fiat lux*: Selective delivery of high flux of nitric oxide (NO) to biological targets using photoactive metal nitrosyls, *Curr. Opin. Chem. Biol.* **2008**, 12, 238–244.

[20] J.E. Clark, P. Naughton, S. Shurey, C.J. Green, T.R. Johnson, B.E. Mann, R. Foresti, R. Motterlini, Cardioprotective actions by a water-soluble carbon monoxide-releasing molecule, *Circ. Res.* **2003**, 93, e2-e8.

[21] T.R. Johnson, B.E. Mann, I.P. Teasdale, H. Adams, R. Foresti, C.J. Green, R. Motterlini, Metal carbonyls as pharmaceuticals? [Ru(CO)$_3$Cl(glycinate)], a CO-releasing molecule with an extensive aqueous solution chemistry, *Dalton Trans.* **2007**, 1500–1508.

[22] I.J.S. Fairlamb, J.M. Lynam, B.E. Moulton, I.E. Taylor, A.-K. Duhme-Klair, P. Sawle, R. Motterlini, η^1-2-Pyrone metal carbonyl complexes as CO-releasing molecules (CO-RMs): A delicate balance between stability and CO liberation, *Dalton Trans.* **2007**, 3603–3605.

[23] J. Niesel, A. Pinto, H.W. Peindy N'Dongo, K. Merz, I. Ott, R. Gust, U. Schatzschneider, Photoinduced CO release, cellular uptake, and cytotoxicity of a tris(pyrazolyl)methane manganese tricarbonyl complex, *Chem. Commun.* **2008**, 1798–1800.

[24] H. Pfeiffer, A. Rojas, J. Niesel, U. Schatzschneider, Sonogashira and "Click" reactions for the *N*-terminal and side chain functionalization of peptides with [Mn(CO)$_3$(tpm)]$^+$-based CO releasing molecules (tpm = tris(pyrazolyl)methane), *Dalton Trans.* **2009**, 4292–4298.

[25] R. Motterlini, P. Sawle, S. Bains, J. Hammad, R. Alberto, R. Foresti, C.J. Green, CORM-A1: A new pharmacologically active carbon monoxide-releasing molecule, *FASEB J.* **2004**, 18, 284–286.

4 Metallointercalators as DNA probes

Summary. The base sequence of DNA encodes all information necessary for an organism to develop. Misregulation of DNA transcription and replication, as well as alteration of coding base sequences are associated with the development of cancer and inheritable human diseases. The specific recognition and manipulation of DNA is thus of great importance for biotechnology and molecular medicine. In this chapter, you will be introduced to the different non-covalent binding modes of transition metal complexes to DNA and the structural features that distinguish them. In the experimental part, you will learn how to prepare such metal complexes and determine the strength of their interaction by UV/Vis and fluorescence spectroscopy.

Learning targets

✓ non-covalent binding modes of small molecules to DNA
✓ interaction of metal complexes with DNA
✓ photophysical properties of ruthenium polypyridyl complexes

Background

The double-helical structure of DNA as determined by Watson and Crick has become one of the iconic images of molecular biology. [1] In the sequence of its nucleobases, all the information necessary to build an organism is encoded.[1] It is thus of paramount importance for the survival of any organism to maintain the integrity of this information. [2, 3] Mutations in coding sequences will lead to proteins with an altered structure and thus an altered activity profile. Modifications to control regions of the genome, on the other hand, will not affect the structure of the protein product itself, but its concentration profile during the life cycle

---

[1] In retroviruses, RNA is the carrier of the genetic information, but reserve transcriptase enzymes transform this to DNA for subsequent mRNA and protein biosynthesis by the host organism.

of a cell. Loss of control of the cell cycle can lead to uncontrolled tissue growth and thus the development of cancer. When such mutations affect germline cells, they can also be passed on to future generations, leading to inheritable diseases.

Transition metal complex binding to DNA has potential applications in both the detection of altered base sequences as well as elimination of cells with acquired mutations. Covalent binding to DNA usually occurs via the nitrogen donor atoms of the nucleobases. This can interfere with DNA transcription and replication, thus leading to cytotoxic effects which can be exploited for cancer chemotherapy. This topic is covered in more detail in Chapter 9. Non-covalent interactions can also lead to cytotoxic effects, but are mostly exploited in bioanalytical applications. Several different binding modes of metal complexes can be distinguished. Before discussing these in detail, the structure of DNA will be briefly recapitulated.

The backbone of the DNA is made up of desoxyribose units linked at their 3'- and 5'-positions via phosphordiester bonds to neighboring nucleosides. In the 1'-position, the desoxyribose is functionalized via an N-glycosidic bond with one of the four planar heteroaromatic nucleobases, adenine (A), guanine (G), cytosine (C), or thymine (T) (Fig. 4.1a). In normal DNA, one of the purine bases always pairs with a pyrimindine base, forming AT or GC base pairs held together by hydrogen bonds (Fig. 4.1b). [4] In the direction of the long axis of the double helix, these Watson–Crick base pairs are positioned on top of each other, with the aromatic ring planes of neighoring nucleobases in a parallel orientation, forming a π stack at the center of the DNA helix. The surface, however, is not uniform but distinctly structured. The sugar-phosphate backbone, which runs in a right-handed way around the long axis of the DNA, leaves two indentions that, due to their different width, are called the major and minor groove, respectively (Fig. 4.2). A total of four different non-covalent DNA binding modes can be distinguished, depending on the type of small molecule binding to the double helix and the recognition elements involved. [5–7]

Electrostatic binding is based on the attractive interaction between the negatively charged phosphate groups of the DNA backbone and cationic ions and metal complexes (Fig. 4.3a). This binding mode is typical for alkali and alkaline earth metal ions like Na^+ and Mg^{2+}, which are responsible for charge compensation in biological buffers. Also, some small globular metal complex cations interact with DNA in this way.

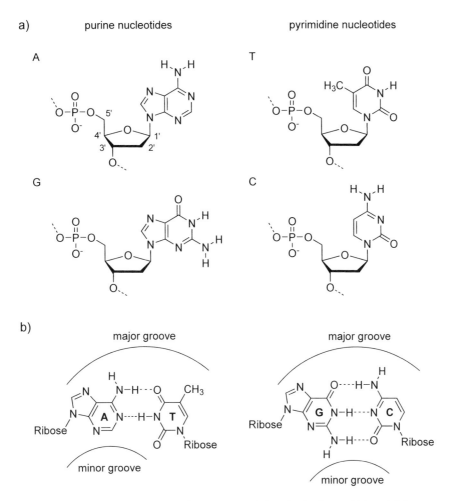

Fig. 4.1 (a) Structure of the four nucleotides A, G, C, and T and (b) orientation of the functional groups of the AT (left) and GC (right) base pairs in the major and minor groove.

Due to the non-directional nature of electrostatic binding, no sequence specificity can be achieved and binding is generally weak.

In the groove-binding mode, molecules sit in either the minor or major groove with close contact to both the functional groups of the sugar-phosphate backbone lining the groove as well as the functional groups from the nucleobases exposed at its bottom. High binding constants can be achieved with linear, extended molecules which wrap around the double helix in one of the grooves and are able to make multiple attractive

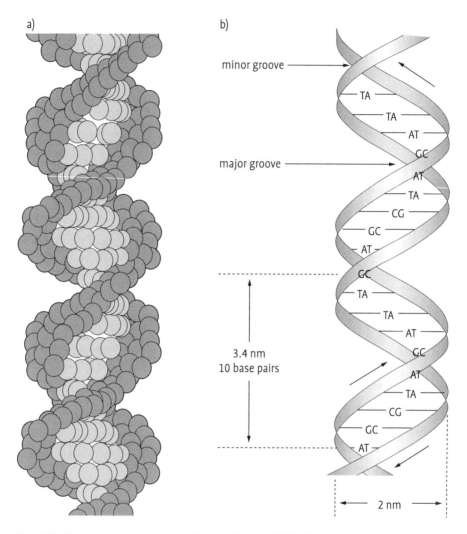

a)

b)

minor groove ———

TA

TA

AT

GC

major groove ———►

AT

TA

CG

GC

AT

GC

TA

TA

AT

3.4 nm
10 base pairs

GC

AT

TA

CG

GC

AT

◄——— 2 nm ———►

Fig. 4.2 Schematic representations of the B-DNA double helix; (**a**) sugar-phosphate backbone in grey and nucleobases in white; (**b**) minor and major groove highlighted together with some metrical parameters of the helical structure.

interactions. Hydrogen bonding to the nucleobase donor/acceptor groups exposed at the bottom of the groove (Fig. 4.2a) leads to especially strong binding and is also the basis of a base sequence-specific recognition which can be achieved with specially-designed groove binders. This is largely the realm of modular extended organic molecules. Some metal complexes are, however, small enough to also bind in one of the grooves (Fig. 4.3b). In

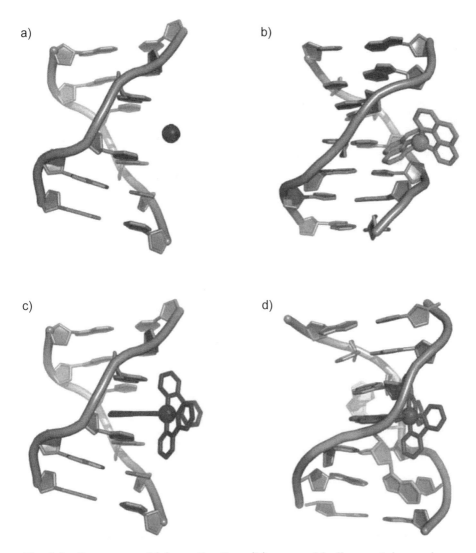

Fig. 4.3 Structures of (**a**) small cation, (**b**) groove-binding metal complex, (**c**) metallointercalator, and d) metalloinsertor bound to a DNA double helix. [B. M. Zeglis, V. C. Pierre, J. K. Barton, *Chem. Commun.* **2007**, 4565–4579] Reproduced by permission of the Royal Society of Chemistry.

chiral octahedral transition metal complexes of the $[Ru(N-N)_3]^{2+}$-type, where N−N is a polypyridyl ligand, different steric interactions of the Λ- and Δ-enantiomers with the rim of the groove form the basis of enantioselective DNA binding observed with these compounds, since the association with the helical-chiral B-form DNA leads to diastereomeric

DNA-metal complex pairs, which thus are different in their physical properties.

The intercalative binding mode is observed for purely organic aromatic molecules as well as metal complexes featuring ligands with an extended planar π-system. In this case, the planar part of the molecule inserts between two Watson–Crick base pairs with the rest of the compound positioned in either the major or minor groove. The backbone of the DNA is expanded to accommodate the intercalator, which acts like an additional base pair inserted into the DNA sequence (Fig. 4.3c). Stacking interactions between the π-systems of the intercalator and the nucleobases give rise to strong binding, which can further be increased by additional weak interactions between the rest of the molecule positioned in the groove and the DNA backbone lining it. These interactions can also enable a recognition of certain base sequences, but since the number of characteristic contacts is usually smaller than with extended groove binders, the specificity is lower and recognition sequences are much shorter. [8]

In some cases, mispairing of nucleobases on opposite strands can occur, either caused by replication errors or chemical modifications of the bases. The mismatches thus formed are significantly destabilized energetically compared to the canonical AT and GC base pairs, since the number of attractive interactions is reduced due to a partial lack of complementarity between hydrogen bond donor and acceptor sites. These mismatches can be recognized by metal complexes which feature special sterically very expansive aromatic ligands. [9] Such ligand systems are too bulky to intercalate into standard DNA and thus show low affinity for oligonucleotide sequences featuring standard Watson–Crick base pairing. At thermodynamically destabilized mismatch sites, however, the expanded ligand can gain access to the DNA base stack and binding in such a position is favored by π-π-interactions with the neighboring aromatic nucleobases. As revealed by the X-ray structure of a rhodium chrysenequinone diimine complex bound to an AC mismatch site, this is accompanied by the two mispaired nucleobases on the opposite strands of the DNA double helix flipped out and replaced by the expanded ligand of the metalloinsertor (Fig. 4.3d). [10] Thus, in contrast to the metallointercalators described in the preceding section, metalloinsertors do not act as a kind of extra base pair leading to an expansion of the DNA base stack, but rather take the place of the mispaired nucleobases in the double helix.

A number of experimental techniques exist to study the interaction of metal complexes with short oligonucleotides as well as longer pieces of DNA. These include NMR and CD (circular dichroism) spectroscopy, mass spectrometry, viscosimetric measurements based on changes in hydrodynamic properties, [11] and electrophoretic techniques as described in Chapter 5. Most commonly employed and easily carried out are, however, methods based on changes in the optical absorption or emission properties of metal complexes free in solution vs. bound to DNA. In the UV/Vis spectral region, the often very intense absorption bands of metal complexes usually undergo a bathochromic and hypochromic shift upon binding to the DNA double helix (Fig. 4.4). While the

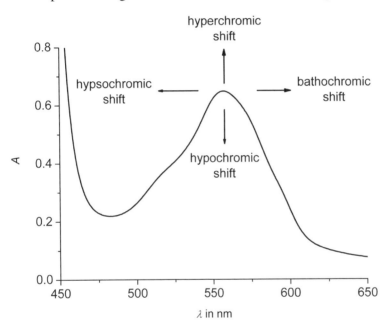

Fig. 4.4 Shifts in band position and intensity possible upon binding of metal complexes to DNA.

origin of the decrease in band intensity is complex, the bathochromic effect (red shift) is thought to be due to electronic coupling between the π stack of the nucleobases and the aromatic ligands. Thus, the more pronounced delocalization of the π-system leads to a decrease in the HOMO-LUMO gap and thus the absorption bands of the metal complex-DNA adduct appear at a lower energy (red shifted) when bound compared to free in solution. A plot of hypochromicity vs. the ratio of c(DNA)/c(metal complex) will then show a saturation behavior upon addition of an increasing amount of DNA to a metal complex solution when it is converted from free in solution to bound to the double helix, given that the DNA affinity is high enough to ensure complete binding. From a more detailed analysis of the absorption changes, the binding constant can also be extracted (see experimental part).

Alternatively, the emission properties of metal complexes can be used to study the DNA binding. Although quite a number of transition metal compounds are fluorescent in anhydrous, degassed, frozen solvents, complexes which emit under physiological conditions are rare. A very important exception is the molecular "light-switch" compound $[Ru(bpy)_2(dppz)]Cl_2$, which is shown in Figure 4.5. In aqueous buffer, this complex is non-fluorescent since the hydrogen bonding of water to the non-coordinated nitrogen atoms of the dppz ligand leads to a thermal deactivation of the excited states. When intercalated into the DNA, the dppz is, however, shielded by the hydrophobic interior of the DNA base stack from such interactions with solvent molecules and thus fluorsecence is observed.

Experiment

Objectives

✓ synthesis of an expanded aromatic polypyridyl ligand and its ruthenium(II) complex
✓ UV/Vis spectroscopy to study DNA binding of metal complexes
✓ determination of hypochromicity and DNA binding constants from UV/Vis titrations

Materials

For metal complex synthesis
– 1,10-phenanthroline ($M = 180.21\,\mathrm{g \cdot mol^{-1}}$)
– potassium bromide ($M = 119.00\,\mathrm{g \cdot mol^{-1}}$)
– concentrated sulfuric acid (96%)
– concentrated nitric acid (65%)
– sodium carbonate/hydrogencarbonate
– anhydrous sodium or magnesium sulfate

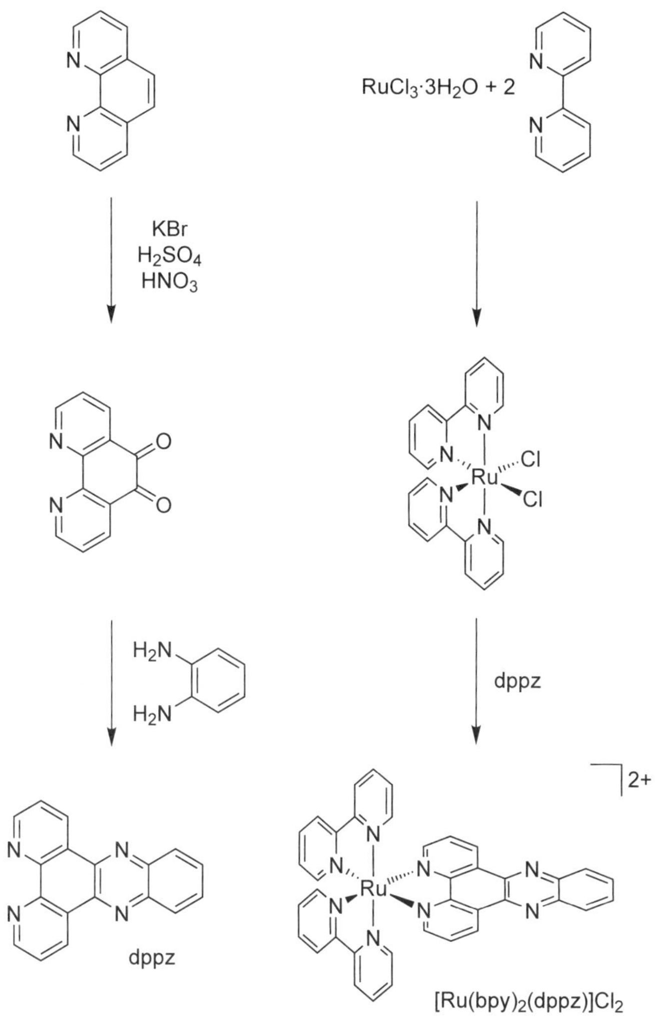

Fig. 4.5 Synthesis of [Ru(bpy)$_2$(dppz)]Cl$_2$ and its precursors.

- 1,2-phenylendiamine ($M = 108.14\,\mathrm{g \cdot mol^{-1}}$)
- 2, 2′-bipyridine ($M = 156.19\,\mathrm{g \cdot mol^{-1}}$)
- ruthenium(III)chloride-trihydrate ($M = 261.47\,\mathrm{g \cdot mol^{-1}}$)
- dimethylformamide
- lithium chloride ($M = 42.39\,\mathrm{g \cdot mol^{-1}}$)
- ammonium hexafluorophosphate ($M = 163.00\,\mathrm{g \cdot mol^{-1}}$)
- tetrabutylammonium chloride ($M = 277.92\,\mathrm{g \cdot mol^{-1}}$)

- ethanol
- acetone

For DNA binding studies

- Tris base, tris(hydroxymethyl)aminomethane, CAS [77-86-1]
- concentrated hydrochloric acid (37%)
- disodium ethylendiaminetetraacetic acid dihydrate
- sodium hydroxide, pellet form
- calf thymus DNA, e. g. Sigma no. D-4522, CAS [91080-16-9]
- ultrapure water
- graded volumetric flasks (10 and 100 ml, 1 l)
- pipettes and tips (10, 100, 1000 μl)
- double-beam UV/Vis spectrometer
- half-micro UV/Vis Quartz cuvettes (750 μl)
- small glass rod to mix the cuvettes
- balance ($\Delta m = 0.1$ mg or better)

Synthesis of metallointercalator [Ru(bpy)$_2$(dppz)]Cl$_2$

Step 1: The dipyridophenazine (dppz) ligand is prepared in a two-step procedure from 1,10-phenanthroline (phen) and 1,2-phenylendiamine. First, phen is oxidized to 1,10-phenanthroline-5,6-dione [12]: in a 250 ml three-neck round-bottom flask equipped with reflux condenser and dropping funnel are placed solid 1,10-phenanthroline (4.0 g) and potassium bromide (4.0 g). The flask is placed in an ice-bath and concentrated sulfuric acid (40 ml) slowly added through the dropping funnel with the acid running down the wall of the flask. The solid materials slowly turn into an orange slurry, which later turns clear red. Then, concentrated nitric acid (20 ml) is added via the dropping funnel. Once the addition is complete, the ice-bath and dropping funnel are removed and the mixture is heated to 100 °C for 3 h. Then, it is poured into 400 ml of ice-cold water in a beaker and left to rest overnight. The yellow solution is carefully neutralized by addition of solid sodium carbonate/hydrogencarbonate (be careful, some foaming may occur, best place the beaker in a plastic tray). Make sure the pH does not rise above 7.0, otherwise undesired byproducts will form. Then extract the aqueous solution with dichloromethane (4 × 100 ml, divide into portions if too large to handle) and leave the combined organic phases standing over anhydrous sodium or magnesium sulfate for at least 1 h, better overnight to remove remaining water. Filter and then evaporate to dryness. Recrystallize the crude product from methanol (take care, the pure 1,10-phenanthroline-5,6-dione is an irritant powder when dry, especially to the respiratory tract). Purity is checked by ^{1}H NMR, IR and mass spectrometry. IR (KBr, cm^{-1}): 1690, 1566, 1462, 1420, 1317, 1300, 1285, 1209, 929, 808, 737; MS (EI): m/z = 210, 182, 154, 127; ^{1}H NMR (CDCl$_3$, ppm): δ = 9.10 (dd, 2H, ^3J = 4.7 Hz, ^4J = 1.8 Hz), 8.48 (dd, 2H, ^3J = 7.7 Hz, ^4J = 2.0 Hz), 7.57 (dd, 2H, ^3J = 7.9 Hz, ^3J = 4.7 Hz).

Step 2: Condensation of the 1,10-phenanthroline-5,6-dione with 1,2-phenylendiamine gives the dppz ligand. In one beaker or Erlenmeyer flask, dissolve 1,10-phenanthroline-5,6-dione (0.34 g) in 15 ml of anhydrous ethanol with heating on a hotplate to obtain a clear yellow solution. In a second flask, dissolve 1,2-phenylendiamine (0.34 g) in the same amount of anhydrous ethanol. Add this solution to the 1,10-phenanthroline-5,6-dione one with a pipette and gently heat for a few minutes. The solution will first turn dirty yellow or brown and then clear up again. Then let cool to room temperature and put in a freezer at +4 °C overnight. The product precipitates in the form of long needles and is collected by filtration. Recrystallize from methanol. IR (KBr, cm^{-1}): 1491, 1416, 1362, 1337, 1074, 816, 761, 739; MS (EI): m/z = 282, 141; ^{1}H NMR (CDCl3, ppm): δ = 9.64 (dd, 2H, ^3J = 8.3 Hz, ^4J = 1.8 Hz), 9.27 (dd, 2H, ^3J = 4.3 Hz, ^4J = 1.8 Hz), 8.35 (dd, 2H, ^3J = 6.5 Hz, ^3J = 3.2 Hz), 7.92 (dd, 2H, ^3J = 6.5 Hz, ^3J = 3.6 Hz), 7.79 (dd, 2H, ^3J = 8.1 Hz, ^3J = 4.5 Hz).

Step 3: Ruthenium trichloride-hydrate (0.60 g) and 2, 2′-bipyridine (0.91 g) are suspended in dimethylformamide (30 ml) and heated to reflux for 3 h. The volume is reduced to one-half and the reaction mixture stored at +4 °C overnight. The precipitated dark solid is collected by filtration, washed with ice-cold water (3 × 25 ml) und dried in the air. The crude product is recrystallized from water (100 ml) containing lithium chloride (1.0 g), precipitated at +4 °C overnight, collected by filtration and air-dried. [13] ^{1}H NMR (dmso-d_6, ppm): δ = 9.97 (d, 2H, ^3J = 4.8 Hz), 8.64 (d, 2H, ^3J = 7.8 Hz), 8.48 (d, 2H, ^3J = 8.0 Hz), 8.04 (m, 2H), 7.63 − 7.80 (m, 4H), 7.51 (d, 2H, ^3J = 5.2 Hz), 7.10 (m, 2H).

Step 4: The final step is the displacement of the two coordinated chloride ligands by dppz. [14, 15] [Ru(bpy)$_2$Cl$_2$]·2 H$_2$O and dipyridophenazine (dppz), 50 mg each, are suspended in a 1 : 1 mixture of ethanol and

water, total volume about 5 ml. The suspension is degassed by passing
through a stream of inert gas (nitrogen or argon) for 10 min. Then, the
mixture is heated to reflux. This can be done either in an oil-bath or in a
microwave reactor (CEM Discover, 120 W, 2 min). The reaction mixture
is then poured into a beaker and water (5 ml) added. It is filtered and a
concentrated solution of ammonium hexafluorophosphate (about 250 mg
in a minimum amount of water) is added to the solution to precipitate the
product as an orange solid, which is collected by filtration and washed with
very little ice-cold water. After drying in the air, the hexafluorophosphate
salt is redissolved in a minimum amount of acetone and a concentrated
solution of tetrabutylammonium chloride (about 250 mg in a minimum
amount of acetone) added to precipitate the product. This is collected
by filtration and dried under vacuum. The purity can be checked by
^{1}H NMR and ESI mass spectrometry. MS (ESI$^+$, water): $m/z = 348.01$;
^{1}H NMR (D$_2$O, ppm): $\delta = 8.98$ (d, 2H, $^3J = 8.0$ Hz), 8.62 (dd, 4H,
$^3J = 8.2$ Hz, $^3J = 4.2$ Hz), 8.23 (dd, 2H, $^3J = 5.2$ Hz, $^4J = 0.8$ Hz),
8.13 (dt, 2H, $^3J = 8.0$ Hz, $^4J = 1.2$ Hz), 8.06 (dt, 2H, $^3J = 8.0$ Hz,
$^4J = 0.8$ Hz), 7.91 − 8.00 (m, 4H), 7.88 (d, 2H, $^3J = 5.2$ Hz), 7.82
(dd, 2H, $^3J = 6.6$ Hz, $^3J = 3.4$ Hz), 7.64 (dd, 2H, $^3J = 8.2$ Hz,
$^3J = 5.4$ Hz), 7.47 (dt, 2H, $^3J = 6.7$ Hz, $^4J = 0.7$ Hz), 7.37 (t, 2H,
$^3J = 6.6$ Hz).

Preparation of 1 M Tris–Cl pH 7.4

In a beaker, 121.14 g of Tris base (tris(hydroxymethyl)aminomethane) is
dissolved in 700 ml of ultrapure water with stirring. After complete dissolution, it is transferred to a 1 l volumetric flask and 60 ml of concentrated
hydrochloric acid are added carefully. After cooling to room temperature,
the pH is checked and adjusted to 7.4 by further addition of small amounts
of concentrated acid if necessary. Then, water is filled up to 1 l. The
solution should be colorless, otherwise the Tris base is contaminated and
should be discarded. [16]

Preparation of 0.5 M EDTA pH 8.0

In a beaker, 186.1 g of disodium ethylendiaminetetraacetic acid dihydrate
(Na$_2$EDTA · 2 H$_2$O) are suspended in 700 ml of ultrapure water under
vigorous stirring. Then, solid sodium hydroxide is carefully added in the
form of pellets to adjust the pH to 8.0. About 20.0 g of sodium hydroxide

are needed. Make sure the pellets have dissolved before adding more. The suspension will only clear up when the pH approaches the desired value. Transfer to a 1 l volumetric flask and fill up with ultrapure water to 1 l. Store the solution in a glass bottle and dilute if needed. [16]

Preparation of 10× TE buffer pH 7.4

In a 1 l volumetric flask, combine 100 ml of 1 M Tris–Cl pH 7.4 and 20 ml of 0.5 M EDTA pH 8.0. Then, add ultrapure water to 1 l and sterilize by filtering or in an autoclave. [16]

Preparation of calf thymus (CT) stock solution

From the original bottle of calf thymus DNA remove the cap and rubber septum. Then, add 4 ml of 1× TE buffer pH 7.4 with a volumetric pipette (do not forget to dilute the 10× stock solution first!) and put septum and cap back in place. Store the bottle in a refrigerator at +4 °C overnight to ensure complete dissolution and resolvation of the DNA. Make sure not to contaminate the DNA with other biological material. Always wear a lab coat and gloves when working with DNA samples. When properly handled and stored, the solution can be used for a week or two. Determine the concentration by UV/Vis spectroscopy as described below.

Preparation of $[Ru(bpy)_2(dppz)]Cl_2$ stock solution

Weigh about 5 mg of the Ru complex and place in a 10 ml volumetric flask. Dissolve by adding 1× TE buffer (do not forget to dilute the 10× stock solution first!) to 10 ml. Determine the concentration by UV/Vis spectroscopy as described below.

Concentration determination of CT DNA and $[Ru(bpy)_2(dppz)]Cl_2$ stock solutions

Take a baseline on the UV/Vis spectrometer using pure 1× TE buffer according to the instructions for your instrument. Then, place the CT DNA or Ru complex stock solution in the sample cuvette and measure a UV/Vis spectrum in the range of 200 to 800 nm for both samples. If necessary, dilute so that the absorption at 260 nm (CT DNA) or 450 nm (Ru complex) is approx. 1.0 and write down the dilution factor. Prepare an additional four dilutions with $0.1 < A < 1.0$ from the respective stock solutions and record their UV/Vis spectra. Then, determine the concentration of these solutions with the aid of the Lambert–Beer law $A = \varepsilon_\lambda \cdot c \cdot l$ and calculate back to the concentration of the stock solutions using $\varepsilon_{260\,nm} = 6600\,l \cdot mol^{-1} \cdot cm^{-1}$ for the CT DNA and $\varepsilon_{450\,nm} = 21\,400\,l \cdot mol^{-1} \cdot cm^{-1}$ for the $[Ru(bpy)_2(dppz)]Cl_2$ complex.

Preparation of CT DNA and Ru complex solutions for the measurement

With the known concentrations of the CT DNA and Ru complex stock solutions, prepare the following three solutions by dilution with an appropriate amount of $1\times$ TE buffer:

- 800 µl of solution A, 10 µM in $[Ru(bpy)_2(dppz)]Cl_2$
- 100 µl of solution B, 10 µM in $[Ru(bpy)_2(dppz)]Cl_2$ and 400 µM in CT DNA
- 100 µl of solution C, 400 µM in CT DNA

UV/Vis titration

Take a new baseline on the UV/Vis spectrometer using pure $1\times$ TE buffer according to the instructions for your instrument. Fill the sample cuvette with 750 µl of solution A and the reference cuvette with 750 µl of $1\times$ TE buffer (volumes refer to half-micro cuvettes). Record a UV/Vis spectrum in the 200 to 800 nm range. Leave both cuvettes in the spectrometer and add to the sample cuvette 5 µl of solution B and to the reference cuvette 5 µl of solution C. Carefully mix with a small glass rod and let equilibrate for 10 min. Then record another UV/Vis spectrum. Repeat the sequence of addition of solutions B and C, mixing, equilibration, and measurement of UV/Vis spectra until three consecutive measurements do not show any spectral changes or the complete 100 µl of solutions B and C have been added.

Evaluation of results

For each addition of metal complex/DNA solutions, calculate the hypochromicity $\%H$ according to eq. (4.1) and plot it against $c(DNA)/c(Ru)$. While the Ru complex concentration remains constant, do not forget to take into account the dilution of the CT DNA. Describe the shape of the curve.

$$\%H = \frac{A_{450\,nm}\,(c_{DNA}=0) - A_{450\,nm}\,(c_{DNA}=x)}{A_{450\,nm}\,(c_{DNA}=0)} \cdot 100 \,. \tag{4.1}$$

Then, determine the intrinsic binding constant K_B according to the method of Schmechel and Crothers. [17] You need to calculate $c(DNA)$ and $c(Ru)$ for each data point and extract the absorption A_λ at a certain peak maximum from the UV/Vis data. In this case, we will use $A_{373\,nm}$, but the evaluation can also be carried out at a different wavelength. Based on the Lambert–Beer law and the law of mass action, K_B is given by eq. (4.2).

$$\frac{1}{\Delta\varepsilon_{ap}} = \frac{1}{\Delta\varepsilon} \cdot \frac{1}{K_B} \cdot \frac{1}{c_{DNA}} + \frac{1}{\Delta\varepsilon} \quad \Leftrightarrow \quad \frac{c_{DNA}}{\Delta\varepsilon_{ap}} = \frac{1}{\Delta\varepsilon} \cdot \frac{1}{K_B} + \frac{1}{\Delta\varepsilon} \cdot c_{DNA} \,. \tag{4.2}$$

In this case, $\Delta\varepsilon = |\varepsilon_b - \varepsilon_f|$ is the difference in extinction between the extinction coefficient ε_b of the fully bound intercalator and ε_f, the extinction coefficient of the free intercalator in the absence of DNA, while $\Delta\varepsilon_{ap} = \varepsilon_a - \varepsilon_f$ is the difference between the apparent extinction coefficient ε_a and ε_f. The apparent extinction coefficient is $\varepsilon_a = \frac{A_{373\,nm}(c_{DNA}=x)}{c_{Ru}}$, which can be obtained from the UV/Vis data for the different DNA concentrations added. Then, determine $A_{373\,nm}(c_{DNA}=0)$ and calculate $\Delta\varepsilon_{ap} = \frac{A_{373\,nm}(c_{DNA}=x)}{c_{Ru}} - \frac{A_{373\,nm}(c_{DNA}=0)}{c_{Ru}}$ for each data point. Next, according to eq. (4.2), plot $c_{DNA}/\Delta\varepsilon_{ap}$ versus c_{DNA}. You should obtain a straight line with a slope of $b = \frac{1}{\Delta\varepsilon}$ and intersection with the ordinate (vertical axis) at $a = \frac{1}{\Delta\varepsilon} \cdot \frac{1}{K_B}$. The two parameters a and b can be determined from a linear fit of the data to $\frac{c_{DNA}}{\Delta\varepsilon_{ap}} = a + b \cdot c_{DNA}$. Then, one obtains $K_B = b/a$.

Variations of the experiment

The procedure described above is also suitable to study the DNA affinity of any other metal complex or organic DNA binder with sufficient solubility in water and significant UV/Vis absorptions bands at $> 300\,nm$. Good results are, however, only obtained in the case of rather high binding constants, of at least 10^4 to $10^5\,M^{-1}$ and above.

There are also more elaborate methods to determine the binding constant K_B which also take into account the width of the binding site and other parameters. These, however, require a non-linear fitting of the absorption data with concentration. [18]

As an alternative to the UV/Vis-based determination of the binding constant, it is also possible to use fluorescence spectroscopy. In the case of luminescent metal compounds like $[Ru(bpy)_2(dppz)]Cl_2$, the emission intensity is simply studied as a function of complex concentration. For this particular complex, the excitation wavelength is 450 nm with the emission best detected at 620 nm. Otherwise, the decrease of ethidium bromide fluorescence can be measured by addition of increasing amounts of intercalating metal complex. In a typical experiment, 100 μM CT DNA and 1 μM of ethidium bromide are used while the metal complex concentration is varied from 0 to 10 μM. [19, 20] Here, an excitation wavelength of 540 nm should be used with the fluorescence detected at 605 nm.

Additional questions

✓ What other forms of DNA exist besides B-DNA? Which structural features distinguish them and under which conditions do they form?

✓ Which different kinds of DNA damage occur in nature and what repair mechanisms are operative to remove them?

✓ Which metals could be used to replace ruthenium and rhodium in the metallointercalators and metalloinsertors described in this chapter?

Bibliography

[1] J. D. Watson, F. H. C. Crick, Molecular structure of nucleic acids – A structure for deoxyribose nucleic acid, *Nature* **1953**, 171, 737–738.

[2] O. D. Schärer, Chemistry and biology of DNA repair, *Angew. Chem. Int. Ed.* **2003**, 42, 2946–2974.

[3] S. Schneider, S. Schorr, T. Carell, Crystal structure analysis of DNA lesion repair and tolerance mechanisms, *Curr. Opin. Struct. Biol.* **2009**, 19, 87–95.

[4] W. Saenger, *Principles of Nucleic Acid Structure*, Springer, New York, **1983**.

[5] K. E. Erkkila, D. T. Odom, J. K. Barton, Recognition and reaction of metallointerca-lators with DNA, *Chem. Rev.* **1999**, 99, 2777–2795.

[6] C. Metcalfe, J. A. Thomas, Kinetically inert transition metal complexes that reversibly bind to DNA, *Chem. Soc. Rev.* **2003**, 32, 215–224.

[7] B. M. Zeglis, V. C. Pierre, J. K. Barton, Metallo-intercalators and metallo-insertors, *Chem. Commun.* **2007**, 4565–4579.

[8] C. L. Kielkopf, K. E. Erkkila, B. P. Hudson, J. K. Barton, D. C. Rees, Structure of a photoactive rhodium complex intercalated into DNA, *Nature Structural Biology* **2000**, 7, 117–121.

[9] B. A. Jackson, J. K. Barton, Recognition of base mismatches in DNA by 5,6-chrysenequinone diimine complexes of rhodium(III): A proposed mechanism for preferential binding in destabilized regions of the double helix, *Biochemistry* **2000**, 39, 6176–6182.

[10] V. C. Pierre, J. T. Kaiser, J. K. Barton, Insights into finding a mismatch through the structure of a mispaired DNA bound by a rhodium intercalator, *Proc. Nat. Acad. Sci.* **2007**, 104, 429–434.

[11] D. Suh, Y.-K. Oh, J. B. Chaires, Determining the binding mode of DNA sequence specific compounds, *Process Biochem.* **2001**, 37, 521–525.

[12] M. Yamada, Y. Tanaka, Y. Yoshimoto, S. Kuroda, I. Shimao, Synthesis and properties of diamino-substituted dipyrido[3,2-a:2′,3′-c]phenazine, *Bull. Chem. Soc. Jpn.* **1992**, 65, 1006–1011.

[13] B. P. Sullivan, D. J. Salmon, T. J. Meyer, Mixed phosphine 2, 2′-bipyridine complexes of ruthenium, *Inorg. Chem.* **1978**, 17, 3334–3341.

[14] E. Amouyal, A. Homsi, J.-C. Chambron, J.-P. Sauvage, Synthesis and study of a mixed-ligand ruthenium(II) complex in its ground and excited states: bis(2, 2′-bipyridine)(dipyrido<3,2-a:2′,3′-c>phenazine-N^4N^5)ruthenium(II), *Dalton Trans.* **1990**, 1841–1845.

[15] Q.-L. Zhang, J.-H. Liu, J.-Z. Liu, P.-X. Zhang, X.-Z. Ren, Y. Liu, Y. Huang, L.-N. Ji, DNA-binding and photoactivated enantiospecific cleavage of chiral polypyridyl ruthenium(II) complexes, *J. Inorg. Biochem.* **2004**, 98, 1405–1412.

[16] J. Sambrook, D. W. Russell, *Molecular Cloning – A Laboratory Manual*, Cold Spring Harbor Laboratory Press, Cold Spring Harbor, NY, **2001**, Vol. 3, Appendix 1.

[17] D. E. V. Schmechel, D. M. Crothers, Kinetic and hydrodynamic studies of the complex of proflavine with poly A · poly U, *Biopolymers* **1971**, 10, 463–480.

[18] J. D. McGhee, P. H. von Hippel, Theoretical aspects of DNA-protein interactions: Co-operative and non-co-operative binding of large ligands to a one-dimensional homoheneous lattice, *J. Mol. Biol.* **1974**, 86, 469–489.

[19] W. C. Tse, D. L. Boger, A fluorescent intercalator displacement assay for establishing DNA binding selectivity and affinity, *Acc. Chem. Res.* **2004**, 37, 61–69.

[20] V. Rajendiran, M. Murali, E. Suresh, M. Palaniandavar, V. S. Periasamy, M. A. Akbarsha, Non-covalent DNA binding and cytotoxicity of certain mixed-ligand ruthenium(II) complexes of 2, 2′-dipyridylamine and diimines, *Dalton Trans.* **2008**, 2157–2170.

5 DNA manipulation with metal complexes

Summary. The manipulation of DNA is important to both biotechnology and cancer chemotherapy. For genomic engineering, it is necessary to cleave DNA strands to remove or introduce oligonucleotide sequences. On the other hand, destruction of the DNA of malignant cells might be beneficial in the chemotherapy of cancer. Among the different methods to study DNA manipulation and degradation, gel electrophoresis is a very simple means to analyze the effect of metal complexes and additives on oligonucleotides. In this chapter, you will learn how to apply this method to study the effect of copper and manganese complexes on plasmid DNA.

Learning targets

✓ different mechanisms of DNA cleavage
✓ forms of plasmid DNA and their migration behaviour through agarose gels
✓ cleavage of DNA by metal complexes

Background

DNA cleavage can occur via two distinctly different processes. The normal way nature achieves this is by hydrolysis of phosphodiester bonds in the backbone of DNA, a reaction that is catalyzed by nuclease enzymes. In this way, 3′- and 5′-ends of the now separate oligonucleotide strands are obtained which can be recognized and reconnected by enzymes of the ligase family. In addition, many nucleases carry out phosphodiester hydrolysis in a base sequence-specific way and there is a significant number of so-called restriction endonucleases with different recognition sequences which are important tools in molecular biology.

On the other hand, modifications of the nucleobases or the desoxyribose sugar unit by redox processes or reactive chemical species can lead to degradation products in which ultimately the phosphodiester bond is also broken. However, such reactions are usually associated with pathological processes or the action of DNA-directed drugs, since such strand breaks cannot easily be connected again by DNA processing enzymes.

Let us first look at enzymatic phosphodiester bond cleavage. Although not strictly a nuclease since it only cleaves off a phosphate group from the end of a nucleotide chain, we will discuss here purple acid phosphatase (PAP) as an example (Fig. 5.1). [1] Since the phosphodiester bond is

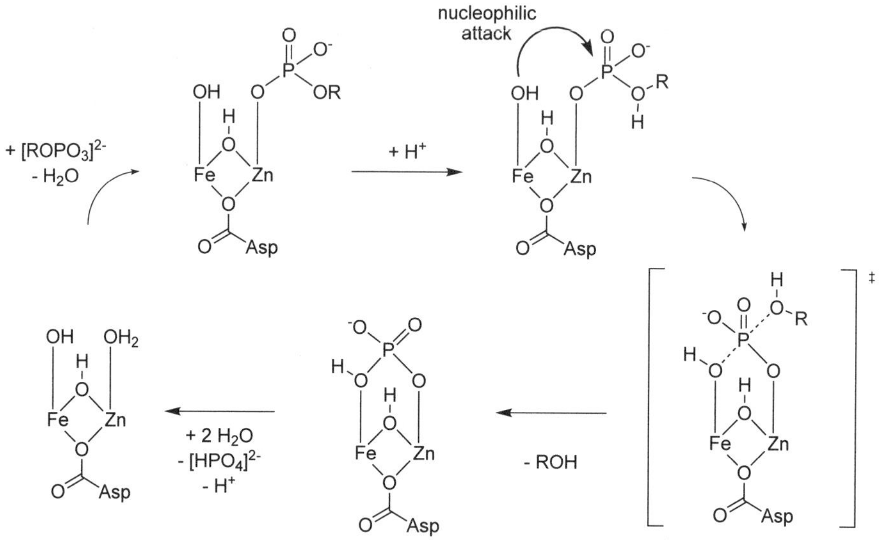

Fig. 5.1 Mechanism of hydrolytic R-O-P bond cleavage by purple acid phosphatase (PAP).

very stable,[1] a base has to be activated for nucleophilic attack at the phosphate group P atom. This is commonly achieved in both enzymes and artificial model compounds by dinuclear metal centers. In the case of PAP, an iron and a zinc center are bridged by the oxygen atom of an aspartate side chain from the protein and a μ-hydroxo ligand. Another OH^--group is terminally bound to the iron center while a water molecule is coordinated to the zinc. The remaining amino acid-derived ligands, four on each metal center, which complete their octahedral coordination sphere, only fix the metal active center and will thus not be considered here any further. In the first step, the water ligand on the zinc center is replaced

---

[1] The half-life of the phosphodiester bonds in DNA has been estimated as 130 000 years under physiological conditions. Phosphodiesterases increase the rate of hydrolytic cleavage by a factor 10^{12}.

by the nucleotide, which coordinates to the metal by an oxygen atom of the $[ROPO_3]^{2-}$ group. Then, the activated terminal hydroxo ligand on the iron does a nucleophilic attack on the P atom of the phosphate group, which is also further activated by the zinc coordination. As the transition state, a five-coordinate phosphorous species is formed and the RO^- group is liberated after protonation as the alcohol. The hydrogenphosphate, however, remains bound to both metal centers in a bridging mode until it is displaced by two water molecules with reformation of the active state, which is then ready for another catalytic cycle.

In order to mimic this reaction in small, artificial nuclease model compounds, usually quite elaborate ligand systems are required to hold a dimetallic active site in the proper geometry for phosphodiester hydrolysis (Fig. 5.2a). [2, 3] Additionally, reaction conditions must be carefully

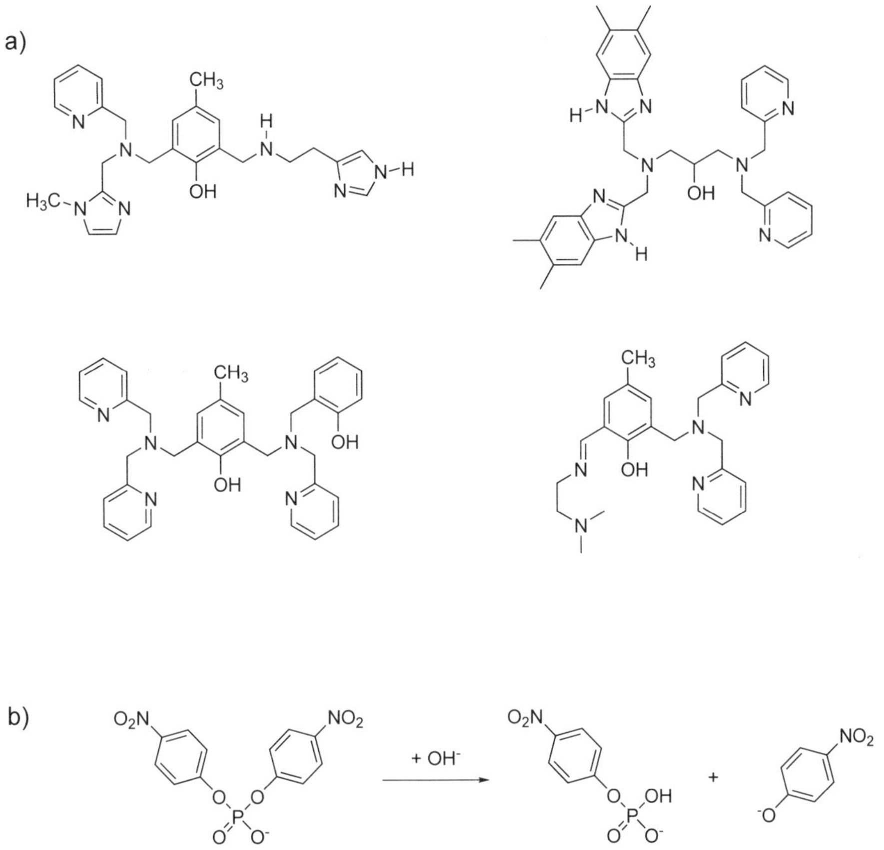

Fig. 5.2 (a) Some examples of ligands used in artificial nuclease model compounds and (b) hydrolytic cleavage of the model substrate 4-BNPP.

optimized for pH value and metal centers used. Although this reaction can very conveniently be followed using UV/Vis spectroscopy on model substrates like bis(4-nitrophenyl)phosphate (4-BNPP, Fig. 5.2b), due to the multi-step ligand synthesis the experiment described here will not touch on such systems.

In this chapter, we will rather focus on oxidative DNA cleavage. A number of metal complexes identified to cleave DNA by such a mechanism are shown in Figure 5.3. Different active species and reactivity either primarily directed at the ribose sugar or the nucleobases can be distinguished. One of the most important cases is probably the reaction of Fe(II) species as in Fe(II)-EDTA with hydrogen peroxide and superoxide, since iron is the most abundant biometal in humans and the two oxygen species are byproducts of the oxygen metabolism of the cell (see Chapter 2). In this so-called Fenton chemistry, ferrous iron (Fe(II)) is oxidized by hydrogen peroxide to generate a hydroxide anion and a hydroxyl radical. The ferric form (Fe(III)) then reacts with superoxide to regenerate to ferrous form (Fe(II)) under production of oxygen.

$$Fe^{2+} + H_2O_2 \longrightarrow Fe^{3+} + {}^{\cdot}OH + OH^-$$
$$Fe^{3+} + O_2^{\cdot -} \longrightarrow Fe^{2+} + O_2$$
$$H_2O_2 + O_2^{\cdot -} \longrightarrow {}^{\cdot}OH + OH^- + O_2$$

In this reaction cycle, a catalytic iron species generates one equivalent of hydroxyl radicals from one molecule of hydrogen peroxide and superoxide each. This highly active and diffusible species then reacts with the nucleobases. Although the lesions generated do not by themselves lead to DNA cleavage, treatment with piperidine and other agents results in fragmentation of the DNA backbone. Such a mechanistic scheme involving hydroxyl radicals can be easily identified by addition of radical scavengers like dimethylsulfoxide, potassium iodide, or mannitol as well as superoxide dismutase or catalase enzymes, which degenerate the reactive oxygen species precursors and thus suppress such reactions. [4] By covalently attaching Fe(II)-EDTA to DNA affinity elements like the organic intercalator propyl methidium (MPE-Fe(II)), such systems can be used in footprinting studies. In such experiments, tight protein binding to the DNA protects certain base sequences from cleavage by the artificial nuclease and thus allows one to determine the binding site. A related chemistry is probably also operative in the biological action of bleomycin, an important iron-based antitumor antibiotic. [5]

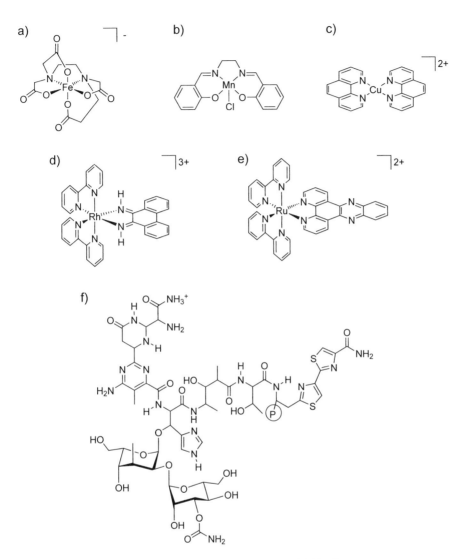

Fig. 5.3 Metal complexes capable of oxidative DNA cleavage:
(a) Fe(II)–EDTA, (b) [Mn(salen)Cl], (c) [Cu(phen)$_2$]$^{2+}$, (d) [Rh(bpy)$_2$(phi)]$^{3+}$,
(e) [Ru(bpy)$_2$(dppz)]$^{2+}$, and (f) bleomycin (the metal-binding site remains a
matter of discussion).

The manganese and copper complexes you will use in this experiment,
which are shown in Figure 5.3b+c, also need hydrogen peroxide for
activation, but no freely diffusible species are formed.   Although the
precise nature of the active compound is not known with certainty, most
likely a high-valent Cu- or Mn-oxo species is formed which then leads to

oxidative damage to the nucleotide followed by cleavage of the glycosidic bond. In subsequent reactions, both the 3'- and 5'-bonds between the phosphorous atoms and the corresponding ribose carbons are cleaved. The 5-methylenefuranone (5-MF) formed leaves a one-nucleotide gap between the disattached 3'- and 5'-ends (Fig. 5.4).

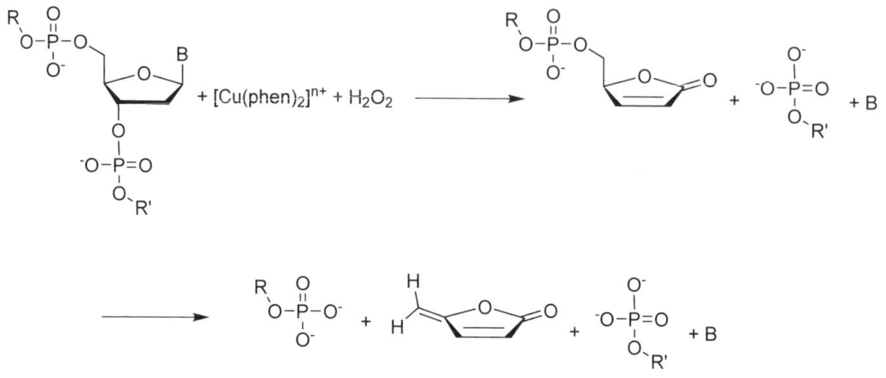

Fig. 5.4 Oxidative reactions leading to cleavage of the phosphodiester bond of DNA (B = nucleobase). The action of the Sigman reagent $[Cu(phen)_2]^{n+}$ is shown here as an example.

In contrast to the iron, manganese, and copper-based systems described above, ruthenium and rhodium complexes, also intensively studied, need light activation to exert their DNA cleavage activity. With Ru(II) compounds, the "flash-quench" technique is commonly employed. Photoexcitation of the ruthenium complex into the metal-to-ligand charge transfer (MLCT) bands lead to a Ru(III)–L$^{\cdot-}$ excited state which is oxidized by an external quencher to a Ru(III)–L species with high oxidative power. If back-transfer from the quencher can be prevented, the Ru(III) complex will oxidize the nucleobases of DNA, with a preference for guanine, which has the lowest oxidation potential. The guanine radical cations thus formed can trigger further interesting reactions, including long-range charge transfer along the DNA double helix. [6–8] In the absence of an external quencher, such photoreactions often lead to the generation of singlet oxygen which will subsequently undergo hydrogen abstraction reactions with the ribose sugar. They can thus be prevented by addition of $^1O_2$-quenchers like sodium azide. The involvement of singlet oxygen can also be identified by carrying out the reaction under an atmosphere of protective gas (dinitrogen or argon), which will prevent such reactions. Replacing water by deuterium oxide (D_2O), on the other hand, will lead to an enhancement of DNA cleavage activity, since the lifetime of singlet oxygen in D_2O is longer compared with that in water.

While it is quite easy for biologists and biochemists to handle genomic DNA, the less experienced researcher interested in the study of oligonucleotide processing needs a well-defined and commercially available form of double-stranded DNA. Very often, plasmids are used for such experiments. A plasmid is a normally circular double-stranded form of extra-chromosomal DNA which is found in bacteria and can be processed by the biochemical machinery of the cell. Since they can undergo autonomous replication in a suitable host, plasmids are important tools in molecular biology to introduce additional genetic material into a cell. The pBR322 plasmid, which will be utilized in the experiment described here, for example, is 4361 base pairs in length and has coding regions for two antibiotics resistance genes (amp^R and tet^R). In addition, there are over forty recognition sequences for restriction enzymes (Fig. 5.5). In our case, we will, however, be less interested in the processing of the plasmid DNA by such enzymes than in the three different forms in which it exists, since they can serve as reporters of DNA modifications by metal complexes as revealed by agarose gel electrophoresis. The commercially available pBR322 plasmid is mostly in the "supercoiled" form (Fig. 5.6a). Here, the DNA is wrapped up to form a very compact structure which travels through the pores of an agarose gel at a speed higher than the corresponding linear double stranded-form. If one single-strand break is induced in one or the other strand of the double-helical plasmid DNA, it unwraps and a circular double-stranded DNA is obtained which is called the "nicked" form (Fig. 5.6b). Due to its large diameter, it passes through the agarose gel matrix at much lower speed than the "supercoiled" form. Finally, if a double-strand break occurs, in which the phosphodiester bonds of the opposite nucleotides in a Watson–Crick base pair are cleaved or two single-strand breaks are only a few bases apart so that the overlapping region is not stable enough thermodynamically to hold the two ends of the circular double-stranded DNA together, the "linear" form is obtained (Fig. 5.6c). Its travel speed is intermediate between the "supercoiled" and the "nicked" form and comparable to that of common DNA length standards, which are also made up of linear pieces of DNA with increasing molecular weight, commonly called a "DNA ladder". If metal complexes form stable covalent or non-covalent adducts with the plasmid, this leads to a reduction of travel speed of any of the three forms due to a combination of increased molecular weight, reduced flexibility, and charge compensation.

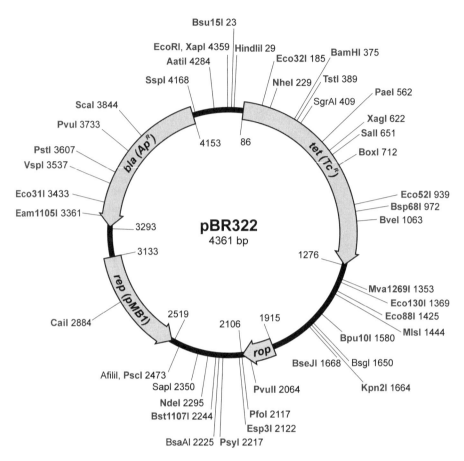

Fig. 5.5 Map of the pBR322 plasmid with the different coding sequences and restriction enzyme cleavage sites shown.

If only one specific adduct is formed, the band simply shifts to lower travel speed while a distribution of adducts of different stoichiometry leads to a smearing of the band (Fig. 5.7). In the case where more than one double-strand break occurs on the plasmid, it is first cleaved in two pieces and then further "chopped up" into a distribution of successively shorter fragments with increasingly higher mobility through the gel, which also becomes apparent as a smear, but at shorter length.

In the present experiment, you will use two different metal complexes based on copper(II) and manganese(II) with 1,10-phenanthroline or salen as the ligand to study oxidative DNA cleavage and investigate the products of the reaction with agarose gel electrophoresis.

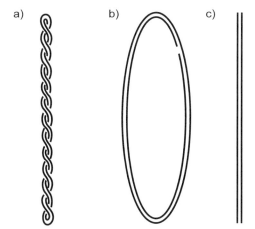

Fig. 5.6 (a) super-coiled, (b) open-circular or nicked, and (c) linear double-stranded form of plasmid DNA.

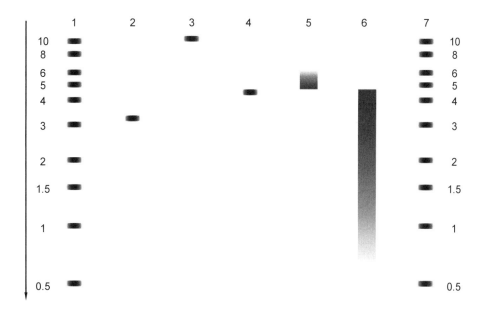

Fig. 5.7 Schematic drawing of an agarose gel showing from left to right: 1) linear DNA length standard, 2) supercoiled plasmid, 3) nicked plasmid, 4) linear plasmid, 5) covalent adducts to linear plasmid, 6) degradation of plasmid to short pieces, and 7) again linear DNA length standard. The numbers to the left and right indicate the length of the DNA length standard in kilobases and the arrow shows the travel direction of the DNA through the gel.

Experiment

Objectives

✓ synthesis of a DNA-cleaving manganese salen complex
✓ investigation of different metal complexes on plasmid cleavage
✓ study of plasmid DNA cleavage by agarose gel electrophoresis

Materials

- salicylaldehyde ($M = 122.12\,\text{g}\cdot\text{mol}^{-1}$)
- 1,2-ethylendiamine ($M = 60.10\,\text{g}\cdot\text{mol}^{-1}$)
- manganese(II)acetate tetrahydrate ($M = 245.09\,\text{g}\cdot\text{mol}^{-1}$)
- potassium chloride ($M = 74.55\,\text{g}\cdot\text{mol}^{-1}$)
- anhydrous ethanol
- ultrapure water
- copper(II)nitrate trihydrate ($M = 241.60\,\text{g}\cdot\text{mol}^{-1}$)
- 1,10-phenanthroline hydrochloride ($M = 216.67\,\text{g}\cdot\text{mol}^{-1}$)
- 30% hydrogenperoxide ($d = 1.11\,\text{g}\cdot\text{ml}^{-1}$)
- pBR322 plasmid (e. g. New England Biolabs N3033S)
- 1 to 10 kbp DNA length standard (e. g. New England Biolabs N3232S)
- 10× TE buffer
- Tris base, tris(hydroxymethyl)aminomethane, CAS [77-86-1]
- disodium ethylendiaminetetraacetic acid dihydrate
- glacial acetic acid
- Ficoll 400 (e. g. Fluka 46327)
- xylene cyanole FF
- bromophenol blue
- agarose (gel electrophoresis grade)
- ethidium bromide
- graded volumetric flasks (10 and 100 ml, 1 l)
- pipettes and tips (10, 100, 1000 µl)
- Eppendorf tubes (1.5 to 2 ml)
- vortexer
- heat block

- small centrifuge
- small (approx. 12×12 cm) horizontal gel electrophoresis tank and power supply
- microwave oven
- three plastic tubs (approx. 14×26 cm)
- UV/Vis lamp and camera or gel documentation system

Synthesis of bis(salicylaldehyde)ethylenediimine (H_2salen)

In a 500 ml three-neck flask fitted with a reflux condenser and a dropping funnel, salicylaldehyde (12.2 g, 100 mmol) is dissolved under argon in anhydrous ethanol (200 ml). Then, the dropping funnel is charged with 1,2-ethylenediamine (3.34 ml, 3.0 g, 50 mmol) dissolved in anhydrous ethanol (30 ml). At room temperature and with vigirous stirring, this solution is added dropwise to the salicylaldehyde. A bright yellow precipitate forms. If this becomes too difficult to stir, another 200 ml of anhydrous ethanol are added to the flask. After complete addition, stirring is continued for another two hours. Then, the precipitated solid is filtered off on a Büchner funnel and dried in the air overnight. Yield: 10.46 g (78%). [9] ^{1}H NMR (dmso-d_6, ppm): $\delta = 13.38$ (s, 2 H, *OH*), 8.59 (s, 2 H, *CH=N*), 7.27 − 7.44 (m, 4 H, phenyl) 6.84 − 6.92 (m, 4 H, phenyl), 3.92 (s, 4 H, N−*CH₂*).

Synthesis of manganese salen complex [Mn(salen)Cl] [9]

In a 100 ml one-neck flask, H_2salen (220 mg) and manganese(II) acetate tetrahydrate (400 mg) are dissolved in ethanol (25 ml). The flask is fitted with a reflux condenser and heated to 110 °C for 3 h. Then, the solvent is completely removed in vacuo. Boiling water (20 ml) is added to the residue and the suspension briefly mixed with a glass rod. It is then poured through a paper filter and the filtrate collected. To this solution, potassium chloride (1.5 g) is added and precipitation of the product completed by storing the flask in a refrigerator at +4 °C overnight. The product is collected on a filter and dried in the air. Due to its paramagnetic nature, no NMR spectrum can be obtained under standard conditions, but the compound can be analyzed with IR spectroscopy and ESI mass spectrometry.

Preparation of 1 M Tris−Cl pH 7.4

In a beaker, 121.14 g of Tris base is dissolved in 700 ml of ultrapure water with stirring. After complete dissolution, it is transferred to a 1 l volumetric flask and 60 ml of concentrated hydrochloric acid added carefully. After cooling to room temperature, the pH is checked and adjusted to 7.4 by further addition of small amounts of concentrated acid if necessary. Then, water is filled up to 1 l. The solution should be colorless, otherwise the Tris base is contaminated and should be discarded. [10]

Preparation of 0.5 M EDTA pH 8.0

In a beaker, 186.1 g of disodium ethylendiaminetetraacetic acid dihydrate ($Na_2EDTA \cdot 2H_2O$) are suspended in 700 ml of ultrapure water under vigorous stirring. Then, solid sodium hydroxide is carefully added in the form of pellets to adjust the pH to 8.0. About 20.0 g of sodium hydroxide are needed. Make sure the pellets have dissolved before adding more. The suspension will only clear up when the pH approaches the desired value. Transfer to a 1 l volumetric flask and fill up with ultrapure water to 1 l. Store the solution in a glass bottle and dilute if needed. [10]

Preparation of 10× TE buffer pH 7.4

In a 1 l volumetric flask, combine 100 ml of 1 M Tris−Cl pH 7.4 and 20 ml of 0.5 M EDTA pH 8.0. Then, add ultrapure water to 1 l and sterilize by filtering or in an autoclave. [10]

Preparation of 50× TAE buffer

In a 1 l volumetric flask, tris(hydroxymethyl)aminomethane (242 g) is suspended in a mixture of 100 ml of 0.5 M EDTA pH 8.0 and 57 ml of glacial acetic acid. Then, about 300 ml of ultrapure water are added and the mixture stirred until all solids are completely dissolved. Finally, ultrapure water is added to 1 l.

Preparation of 6× gel loading buffer

In a 250 ml beaker, 7.5 g of Ficoll 400 are suspended in 50 ml of ultrapure water and the mixture is stirred for 1 h until a clear, colorless solution is obtained. Then, 125 mg of Xylene cyanole FF and 125 mg of Bromophenol blue are added and the deep blue highly viscous solution is stirred for another 15 min. It is then transferred to a brown screw-capped bottle and should be stored at +4 °C in the dark. A change in color indicates a degradation of the gel loading buffer and it should then be discarded.

Preparation of ethidium bromide stock solution

Safety note: Ethidium bromide avidly binds to DNA, is highly staining and toxic to humans. Make sure not to spill anything and protect your skin from contact with the material. Wear a lab coat and two pairs of gloves. Pick up any spilled materials with paper and dispose of properly. In a 50 ml beaker, place 10 ml of ultrapure water and add 10.7 mg of ethidium bromide. Dissolve by stirring at room temperature protected from light for a few hours. Then transfer to a brown screw-capped bottle and store at room temperature protected from light, for example by wrapping in aluminum foil.

Preparation of stock solutions for DNA cleavage

All the following stock solutions should be prepared freshly before the start of the experiment.

Prepare a 2 mM copper(II) solution by putting about 10 mg of copper(II) nitrate trihydrate in a 10 ml volumetric flask and fill up to the mark with 1× TE buffer. Calculate the concentration of the solution from the volume of buffer and amount of copper salt added. Dilute accordingly if necessary.

Prepare a 4 mM phen solution by putting about 10 mg of 1,10-phenanthroline hydrochloride in a 10 ml volumetric flaks and fill up to the mark with 1× TE buffer. Calculate the concentration of the solution from the volume of buffer and amount of copper salt added. Dilute accordingly if necessary.

Prepare a 1 M solution of hydrogen peroxide. In a small plastic vial, place 898 μl 1× TE buffer. Then, add 102 μl of 30% hydrogen peroxide and mix on a vortexer. Then, dilute 1 : 1000 to also prepare a 1 mM solution of hydrogen peroxide. If you do not have pipettes of the required accuracy around, use three serial 1 : 10 dilutions instead.

Prepare a 2 mM solution of [Mn^{III}(salen)Cl] by putting about 10 mg of the manganese(II) salen complex in a 10 ml volumetric flaks and fill up to the mark with 1× TE buffer. Calculate the concentration of the

Table 5.1 Pipetting scheme for the DNA agarose experiment.

	sample 1	sample 2	sample 3	sample 4	sample 5
copper(II)nitrat trihydrate c_{stock}	–	2 mM	2 mM	–	–
V_{stock}	–	1.0 µl	1.0 µl	–	–
c_{final}	–	40 µM	40 µM	–	–
1,10-phenanthroline c_{stock}	–	4 mM	4 mM	–	–
V_{stock}	–	2.0 µl	2.0 µl	–	–
c_{final}	–	160 µM	160 µM	–	–
[Mn(salen)Cl] c_{stock}	–	–	–	2 mM	2 mM
V_{stock}	–	–	–	10 µl	1.0 µl
c_{final}	–	–	–	400 µM	40 µM
30% hydrogen peroxide c_{stock}	–	1 mM	1 mM	1 M	1 M
V_{stock}	–	20.0 µl	2.0 µl	2.0 µl	2.0 µl
c_{final}	–	0.4 mM	0.04 mM	40 mM	40 mM
pBR322 plasmid DNA c_{stock}	1.5 mM	1.5 mM	1.5 mM	1.5 mM	1.5 mM
V_{stock}	1.3 µl	1.3 µl	1.3 µl	1.3 µl	1.3 µl
c_{final}	40 µM	40 µM	40 µM	40 µM	40 µM
1× TE buffer V_{stock}	48.7 µl	25.7 µl	43.7 µl	36.7 µl	45.7 µl
V_{total}	50.0 µl	50.0 µl	50.0 µl	50.0 µl	50.0 µl

solution from the volume of buffer and amount of copper salt added. Dilute accordingly if necessary.

DNA cleavage experiments

According to the concentrations given in Table 5.1, prepare five small plastic vials with your samples by pipetting in the stock solutions as indicated in the following order: a) buffer, b) plasmid DNA, c) metal complex and ligand, and d) hydrogen peroxide. If necessary, appropriately dilute your stock solutions first. Then, vortex all vials to ensure mixing, briefly spin down with the centrifuge and incubate for 1 h at 37 °C in the heat block. During this time, protect the vials from light by covering with aluminum foil. During the incubation time, proceed with preparation of the agarose gel.

Preparation of 1% agarose gel

Make sure that the bottom of your gel chamber is adjusted parallel to the bench level and the comb (14 wells) is at hand. Place 0.5 g of electrophoresis grade agarose in an Erlenmeyer flask and suspend by addition of 50 ml 1× TAE buffer. Heat the suspension in a commercial microwave oven at full power (700 W) for about three minutes until it clears up. Very carefully remove from the microwave since the solution and flask are very hot and may boil over due to superheating. Make sure to only gently swirl the solution to prevent the formation of bubbles. Let cool to about 70 °C since higher temperature might damage the plastic of the gel chamber. Then carefully pour the gel into the chamber to make sure to get an even coverage. If some small bubbles appear, they can be removed with the aid of a Pasteur pipette. Do not forget to put in the comb while the gel is still hot. Cover the gel chamber and let it solidify for approx. 1 h. Then adjust the gel electrophoresis setup in such a way that the comb is towards the negative pole and facing the edge of the bench while the positive pole is towards the back. Carefully pour 1× TAE buffer into the gel chamber so that the gel is covered by 2 to 3 mm of buffer. Then, very carefully remove the comb.

Preparation of samples for loading to the gel

From each sample vial, take 10 μl and transfer them to another empty vial. Then, add 2 μl of 6× gel loading buffer to each new vial. In addition, prepare two vials with 1 to 10 kbp DNA length standard by first adding 9 μl of 1× TE Buffer followed by 1 μl of 1 kbp DNA ladder and 2 μl of 6× gel loading buffer. Thoroughly mix each vial on the vortexer and then briefly spin down with the centrifuge.

Loading of the gel

Load 8 μl of each mixture of sample and gel loading buffer to the wells. Place the DNA ladder samples to the very left and right of the gel as shown in Figure 5.7. Make sure not to damage the bottom of the wells. Especially if you are inexperienced with gel electrophoresis, load samples only to every other wells to prevent cross-contamination. Work carefully but quickly to make sure the loading buffer does not diffuse out of the gel. Carefully note which sample has been loaded to which well.

Running the gel

Make sure the power supply is connected properly to the gel tank. The negative pole should be attached to the side where the wells are located and the positive one to the opposite side (why?). Then run the gel for about 2 h at 100 V. Check occasional if the gel is still running correctly. The loading dye will separate in two colored bands, one blue and one magenta. Stop the electrophoresis before the more quickly moving dye band reaches the end of the gel.

Staining of the gel

Safety note: Be careful – ethidium bromide avidly binds to DNA and is toxic to humans. Make sure not to spill any solution and protect your skin from contact. Wear a lab coat and two pairs of gloves. Afterwards, make sure to properly dispose off the ethidium bromide solutions according to the regulations of your institution. Do not pour into the sink!!!

Fill two plastic tubs about twice the size of your gel with 1 l of ultrapure water each. To the first one, add 50 µl of ethidium bromide stock solution and mix with a glass rod. Take the gel from the electrophoresis tank and soak in the ethidum bromide solution for 10 min. Then remove the gel and place in the plastics tub with pure water for another 10 min to remove unbound ethidium bromide. Transfer the gel to a suitable container and take a picture of it with a combination of UV/Vis lamp and camera or a gel documentation system. The orange fluorescence of the DNA-bound ethidium bromide should be clearly detectable even by visual inspection under UV excitation. **Safety note:** Make sure to protect your eyes and skin from UV light and properly dispose of the gel after taking the picture. Do not dispose in the normal trash!

Evaluation of results

Load the picture of the gel into a suitable graphics program and label the wells. Rotate the picture so that the wells are at the top. First, identify the bands of the DNA ladder standard to the left and right of the gel. Then, compare the number of bands and their intensity for the different incubation conditions in a qualitative way. If possible, quantify the intensity of each band and compare relative to a standard. What do the changes in relative band intensities mean?

Variations of the experiment

The procedure described above can be used to follow plasmid DNA modification by other metal complexes too. For example, use the ruthenium(II) metallointercalator $[Ru(bpy)_2(dppz)]Cl_2$ from Chapter 4 to study light-induced DNA cleavage or try the Fenton chemistry of the Fe(II)-EDTA system. You might also want to study the influence of concentration of metal complexes and additives, incubation temperature and time, or light on the DNA cleavage. In the case of strongly DNA intercalating compounds (see Chapter 4) be aware of the fact the these might remain bound to the plasmid DNA even during incubation, gel electrophoresis, and staining so that the ethidium bromide cannot bind to the DNA double helix and thus the fluorescence signal is either very weak or absent, even though the DNA has not been degraded (false negative result).

Additional questions

✓ What other enzymes than PAP are known to hydrolytically cleave DNA?

✓ How do restriction enzymes achieve a sequence-specific cleavage? Do they need metal active sites to function?

✓ Why is 4-BNPP such a good model substrate for phophordiester hydrolysis?

Bibliography

[1] G. Parkin, Synthetic Analogues relevant to the structure and function of zinc enzymes, *Chem. Rev.* **2004**, 104, 699–767.

[2] A. Neves, M. Lanznaster, A. J. Bortoluzzi, R. A. Peralta, A. Casellato, E. E. Castellano, P. Herrald, M. J. Riley, G. Schenk, An unprecedented $Fe^{III}(\mu\text{-}OH)Zn^{II}$ complex that mimics the structural and functional properties of purple acid phosphatases, *J. Am. Chem. Soc.* **2007**, 129, 7486–7487.

[3] A. Greatti, M. Scarpellini, R. A. Peralta, A. Casellato, A. J. Bortoluzzi, F. R. Xavier, R. Jovito, M. Q. de Brito, B. Szpoganicz, Z. Tomkowicz, M. Rams, W. Haase, A. Neves, Synthesis, structure, and physicochemical properties of dinuclear Ni^{II} complexes as highly efficient functional models of phosphohydrolases, *Inorg. Chem.* **2008**, 47, 1107–1119.

[4] D. S. Sigman, A. Mazumder, D. M. Perrin, Chemical nucleases, *Chem. Rev.* **1993**, 93, 2295–2316.

[5] R. M. Burger, Cleavage of nucleic acids by bleomycin, *Chem. Rev.* **1998**, 98, 1153–1169.

[6] R. E. Holmlin, P. J. Dandliker, J. K. Barton, Charge transport through the DNA base stack, *Angew. Chem. Int. Ed.* **1997**, 36, 2714–2730.

[7] S. Delaney, J. K. Barton, Long-range DNA charge transport, *J. Org. Chem.* **2003**, 68, 6475–6483.

[8] E. J. Merino, A. K. Boal, J. K. Barton, Biological contexts for DNA charge transport chemistry, *Curr. Opin. Chem. Biol.* **2008**, 12, 229–237.

[9] S. R. Doctrow, K. Huffman, C. B. Marcus, G. Tocco, E. Malfroy, C. A. Adinolfi, H. Kruk, K. Baker, N. Lazarowych, J. Mascarenhas, B. Malfroy, Salen-manganese complexes as catalytic scavengers of hydrogen peroxide and cytoprotective agents: Structure-activity relationship studies, *J. Med. Chem.* **2002**, 45, 4549–4558.

[10] J. Sambrook, D. W. Russell, *Molecular Cloning – A Laboratory Manual*, Cold Spring Harbor Laboratory Press, Cold Spring Harbor, NY, **2001**, Vol. 3, Appendix 1.

6 Synthesis of metal-peptide bioconjugates

Summary. This chapter will familiarize you with organometallic bioconjugates in the form of metallocene peptides. These compounds will be prepared by solid phase peptide synthesis (SPPS), which is the most common and powerful technique for the synthesis of small to medium-sized peptides. You will be introduced to the principles of SPPS methods and the considerations and precautions necessary to incorporate metal complexes. In the experimental part, you will perform the SPPS of a neuropeptide and attach metal complexes to it, isolate and purify the metal-peptide product, and perform common characterization experiments such as chromatography and mass spectrometry.

Learning targets

✓ Chemical structure of peptides
✓ Chemical synthesis of peptides by solid phase peptide synthesis (SPPS) techniques
✓ Introduction of metal complexes, in particular metallocenes, as part of the SPPS scheme
✓ Chromatographic analysis and purification of peptides and their metal conjugates
✓ Spectroscopic characterization of metal-peptide conjugates (see also Chapter 9)

Background

In this experiment, we prepare a naturally occurring biomolecule (a peptide) functionalized with a metal complex. Such a combination is commonly denoted a metal bioconjugate. More specifically, we use organometallic complexes, i. e. metal complexes with at least one direct, covalent metal-carbon bond. The compounds described herein thus constitute organometal-peptide bioconjugates, and the area of research that deals with such compounds is nowadays commonly defined as *bioorganometallic chemistry* – a part of bioinorganic chemistry which specifically uses organometallic compounds. [1, 2] Traditionally, bioorganometallic chemistry describes a somewhat odd combination of fields. Organometallic compounds, although having multiple applications in material science and catalysis, are perceived by many chemists as being sensitive towards oxygen and moisture. They were thus for a long time deemed incompatible with biological applications, which usually require water as solvent and working in air. This, of course, turned out to be a biased view and today there is indeed a surprising variety of organometallic compounds which are perfectly compatible with biochemical requirements. Organometallic complexes possess special properties, which can be used, inter alia, for the selective and highly sensitive detection of biomolecules. Examples of such methods would be electrochemical techniques (see also Chapter 8), and infrared (IR) spectroscopy of metal carbonyl complexes, which has been developed into a versatile immuno assay. [3–5] Alternatively, the metal complex can be quantified directly by atomic absorption spectroscopy (AAS) and this can provide valuable insight into cellular processes modulated by bioactive metal compounds or metal bioconjugates. [6, 7] In this experiment, we perform the chemical synthesis of a peptide to which an organometallic compound is covalently attached. [8, 9] Such organometallic peptide conjugates have been proposed as new antibiotics, [10] to investigate the mechanism of action of anti-cancer drug candidates, [11] or to follow the uptake and intra-cellular localization of the attached metal complexes. [6, 12, 13]

Amino acids are the natural building blocks for many large biopolymers, in particular peptides and proteins (see Fig. 6.1). Both consist of a linear chain of amino acids. Commonly, peptides are small oligomers with less than about 30 amino acids (some people prefer to say less than 50 amino acids), whereas proteins are larger, with molecular weights between

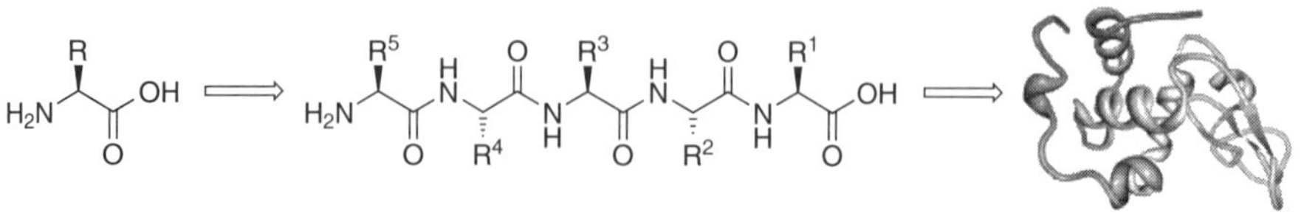

Fig. 6.1 From amino acids (left) to peptides (a pentapeptide, middle) to proteins (right, schematic structure of lysozyme, PDB code 133I).

ca. 10 kDa and up to several hundred kDa. Methods for the chemical modification of full-size proteins will be treated in the following Chapter 7. Herein, we deal with peptides only, which are readily synthesized and hence also easily modified in the chemical laboratory.

Peptides have a variety of biological functions. They often serve as signaling or mediator molecules. They are naturally "produced" by cleavage from larger proteins, or even by fragmentation of a protein into several peptides. Proteins, on the other hand, are produced by a series of events on the ribosome, which is essentially an assembly of proteins and RNA molecules that constitute the protein synthesis machinery. As described in Chapter 7, the DNA sequence (or gene) ultimately controls the composition and sequence of amino acids in a protein. Therefore, modification of the amino acid sequence in a protein (and hence its cleavage product, a peptide) requires changing the DNA of that specific protein. Although such procedures are fairly standard nowadays in a molecular biology laboratory, it is still a cumbersome task and there is no guarantee for success in all cases. Along the same lines of argument, it is even more difficult to introduce non-natural amino acids, because there is evidently no gene that will code for them, let alone the necessary molecular machinery to include them into proteins.

Therefore, the chemical synthesis of peptides has changed modern biology tremendously. Not only are chemical changes in the amino acid sequence easy to achieve, but also the introduction of an almost unlimited number of non-natural amino acids and functionalities is in most cases as easy as simply adding another amino acid. Peptides are today synthesized by a series of steps jointly called solid phase peptide synthesis (SPPS). This procedure was developed in the 1960s by Bruce Merrifield, [14] who was awarded the Nobel Price in Chemistry for his invention in 1984. An SPPS scheme is depicted in Figure 6.2. [8] Principally, SPPS consists of repeated cycles of N-terminal protecting group removal on the last amino acid, which is immobilized on the solid support, and coupling of the next incoming amino acid. The new amino acid has to be activated to ensure quick and smooth peptide coupling, and all side chains have to be chemically protected so as to avoid reactions in undesired places.

SPPS is a sequential synthesis of linear peptides from amino acid building blocks on a solid support, usually called the resin. The resin serves as a support on which the peptide chain grows. It is usually an organic polymer with some mechanical stability as the resin is agitated during synthesis. At the start of the synthesis, the resin must be swollen

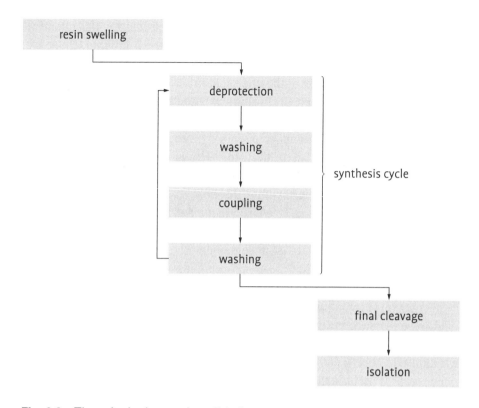

Fig. 6.2 The principal steps in solid phase peptide synthesis (SPPS).

in the solvent of choice to ensure that all reactive groups are exposed to the reagents which should diffuse freely inside the resin. Polar, non-protic solvents such as DMF or NMP (*N*-methyl-pyrrolidine) are chosen in which all reagents should be reasonably soluble. The growing peptide chain is covalently connected to the resin by a so-called "linker". The linker is chemically stable during SPPS, but labile under certain conditions. At the end of the SPPS scheme, the peptide is cleaved from the resin by exposing it to exactly those conditions. Unless a protected peptide is required for further chemical reactions, all side chain protecting groups are removed at the time of cleavage as well. The original Merrifield resin and linker required very harsh acidic (conc. hydrofluoric acid) conditions for cleavage. Standard resins used today still require acid cleavage, but less harsh conditions. Typically, 90% TFA (trifluoroacetic acid), 5% water and 5% tri(isopropyl)silane (TIS) are used. The silane is added to trap undesired radicals and carbocations which are generated during the cleavage process. It is a most convenient choice (safe, odorless, volatile), but can be replaced by thiols, phenols, or related agents. Beyond that, a variety of linkers are available for acid and base cleavage at different concentrations, cleavage by nucleophiles to produce *C*-terminally functionalized peptides, or even cleavage by redox reactions or light. This makes the chemical synthesis of peptides by SPPS methods tremendously versatile, and it even allows the incorporation of sensitive metal complexes, provided that a prudent choice of resin, linker, and protecting groups is made at the planning of the synthesis. In this experiment, we use the Wang resin, which is a polystyrene resin with a benzyloxy-benzyl alcohol linker as depicted in Figure 6.3. [15,16]

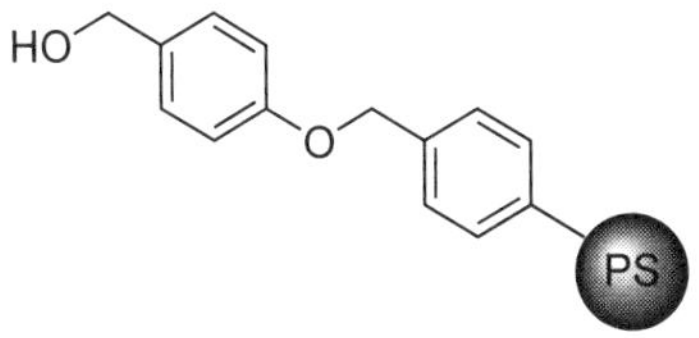

Fig. 6.3 Wang resin: *p*-Benzyloxy-benzyl alcohol linker on polystyrene (PS).

In order to achieve rapid and complete peptide coupling, the incoming amino acid must be activated. To this end, a variety of so-called coupling reagents is available. Traditionally, diimides such as the famous dicyclohexylcarbodiimide (DCC) were used. Although they still have a role in peptide coupling in solution, only moderate activation and problems removing side products have largely led to replacement of the carbodiimides. Popular reagents nowadays include phosphonium (BOP, pyBOP) or uronium salts (HBTU, TBTU, HATU) (see Figure 6.4 and list of abbreviations). Both HBTU and TBTU are salts of the same cation. They only differ in their anion (PF_6^- in HBTU and BF_4^- in TBTU), which accounts for slightly different solubility but very similar coupling efficiency. A review of all common coupling reagents was written by Montalbetti and Falque, [17] and a critical review appeared recently. [18] In this experiment, we use TBTU throughout. [19] The amino acids are pre-activated for some time before they are added to the resin. This ensures formation of the activated ester which is ready to react with the free amino groups of the peptide chains. To speed up the formation of activated esters, we add hydroxybenzotriazole (HOBt) to the preactivation mixture. [20] Rapid formation of the activated esters and smooth peptide

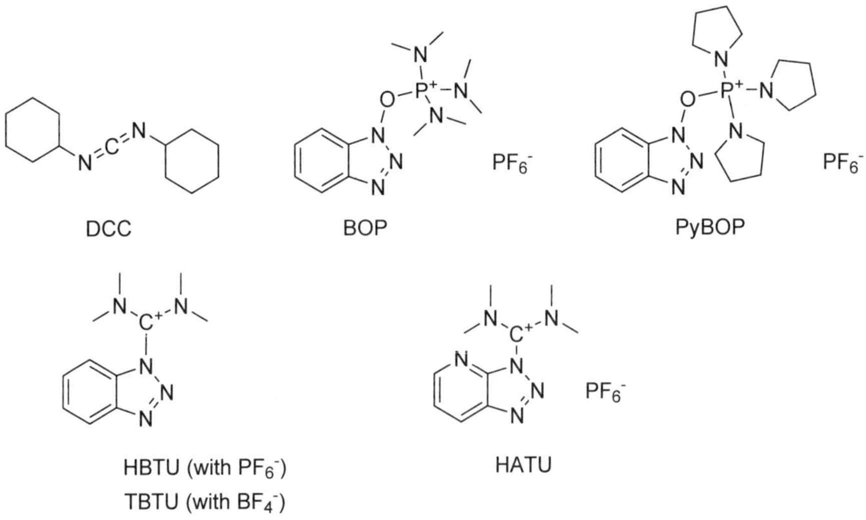

Fig. 6.4 Chemical structures of commonly used SPPS coupling reagents.

bond formation will also decrease the risk of racemization of the amino acids. Naturally occurring peptides consist almost exclusively of L amino acids, which is the natural enantiomer. In the chemical synthesis of peptides by SPPS, there is no restriction to the chirality of the amino acid – another advantage of chemical synthesis. Very often, one or a few D amino acids are introduced into peptides for medicinal or cell biology applications to enhance their stability towards proteolytic degradation by peptidases.

In order to make the coupling reaction specific, the amino terminus of each new amino acid is protected by a temporary protecting group. The two main protecting groups are *tert*-butoxycarbonyl (Boc) and fluorenyl-methoxycarbonyl (Fmoc) (see Figure 6.5). The Boc group is cleaved by strong acid, usually conc. trifluoroacetic acid (TFA), and is also used as a permanent protecting group for some side chains such as the amino group in lysine. For most peptide syntheses today, the Boc group has been replaced by the more versatile Fmoc group. [21] The Fmoc group is cleaved by mild bases (usually 20% piperidine in DMF). This leaves room for a variety of side chain protecting groups with graduated reactivity, as long as they are stable to the basic conditions of Fmoc removal. We use Fmoc chemistry in this lab project throughout.

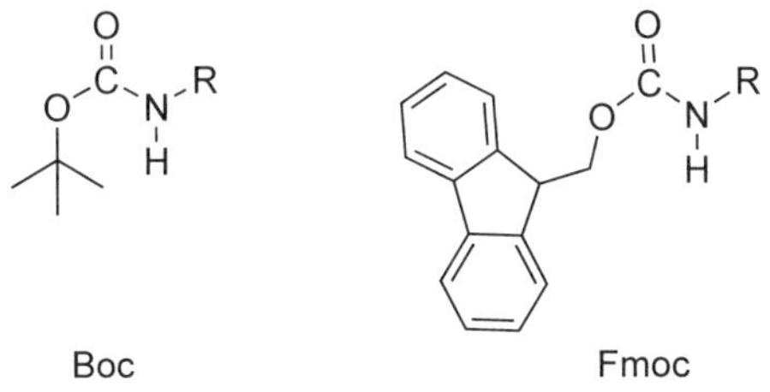

Fig. 6.5 The two most common N-terminal SPPS protecting groups.

The last question to address are reactive side chains such as the thiol group in cysteine, the guanidinium group in arginine, the amino groups in lysine and histidine, carboxylates in glutamic or aspartic acid, or hydroxy groups in serine, threonine or tyrosine. These groups are protected throughout the whole SPPS process by permanent (or non-temporary) protecting groups. The permanent protecting groups are only removed after the whole peptide is assembled during cleavage of the peptide from the resin. Alternatively, permanent protecting groups may be chosen such that they can be removed selectively by specific reagents and chemistry. This will release only the one side chain that was deprotected, which can then be further functionalized while all other amino acids of the peptide remain protected. In this way, only one specific lysine can be functionalized out of a poly-lysine sequence, for example. To describe this, organic chemists speak of orthogonal protecting groups. Orthogonal in this context means that one specific protecting group can selectively be removed without affecting all other groups. For example, the combination of Fmoc amino acids with acid-labile side chain protecting groups is an orthogonal combination of temporary and permanent protecting groups. To add the chloro-trityl group, which can be removed from the tyrosine phenol group by dilute acid (e. g. 10% TFA in DMF) would introduce yet another orthogonal protecting group. To add more complexity, a glutamic acid could be protected by an allyl ester, which is selectively removed by zero-valent Pd complexes. Evidently, this offers literally millions of possible combinations for selective and multiple derivatizations of a peptide. Nowadays, a large number of amino acids with almost all possible combinations of protecting groups, as well as different resins with linkers are commercially available. Clearly, it is absolutely essential to plan all details well before starting the SPP synthesis of a long and complicated peptide. If any doubt remains about the required reactivity, test reactions on single amino acids or small peptide fragments should be performed to ensure that the planned combination of chemical functionalities will perform as desired. This is particularly true if "uncommon" compounds come into play – such as the metal complexes that we describe in this experiment. A complete overview of protecting groups, not just for peptide synthesis, can be found in the manual *Protective Groups in Organic Synthesis*. [22]

For this experiment, we have chosen a fairly short and easy to synthesize peptide, which will nevertheless demonstrate the principle of SPPS and the associated considerations. Enkephalin is a five amino acid neuropeptide. It was first isolated in 1975 from mammalian brains and has pronounced analgesic effects. [23] Two forms of enkephalin exist with the primay amino acid sequences Tyr-Gly-Gly-Phe-Leu ([Leu]^5-enkephalin) and Tyr-Gly-Gly-Phe-Met ([Met]^5-enkephalin, both sequences are written from N- to C-terminus). The two enkephalins are cleaved from a larger precursor protein called pre-enkephalin which is produced by protein biosynthesis directly inside neuronal cells. Enkephalins are natural ligands for the opiate receptor, which explains their analgesic effects in the brain. However, like many other neuropeptides, they possess multiple functions in other organs. In fact, they seem to be fairly ubiquitous in the body, although at low concentrations and under tight control. We concentrate on the more common [Leu]^5-enkephalin in this experiment, which is often abbreviated as "Enk".

In this experiment, we integrate organometal complexes into enkephalin during the SPPS scheme. [9] As described above, metal complexes may pose particular problems to otherwise straight-forward (organic) synthetic procedures. This is particularly true for organometallic complexes, which are frequently low-valent (and hence prone to oxidation) and may have hydrolytically labile metal-carbon bonds. In this experiment, we use the metallocenes ferrocene carboxylic acid, cobaltocenium carboxylic acid und ruthenocene carboxylic acid as metal complexes (Figure 6.6). They are attached to the peptide in the last step of the SPPS while the peptide is

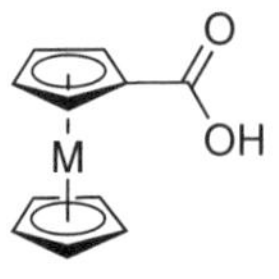

$M = Fe, Co^+, Ru$

Fig. 6.6 Metallocene carboxylic acids used in this experiment.

still on the resin. To this end, the carboxylic acid is activated with TBTU as if it were an amino acid. However, the coupling proceeds slower for the cobaltocenium acid, and therefore reaction overnight is required. [12, 13] If the activity of a novel (amino) acid is not known, then a small-scale reaction should be made with just a one or two amino acid peptide that mimics the site of attachment to the peptide. A few beads of the resin can be removed after a given time period, the peptide is cleaved from those beads, precipitated with ether, and the residue directly analyzed by HPLC and/or LC-MS. The reaction is continued and the procedure repeated until complete conversion is indicated by pure chromatograms and/or mass spectra.

Although all three metallocenes are formally stable 18-electron complexes in preferred, non-zero oxidation states of the metals (see Chapter 9 for a description of their electronic structure), their incorporation into the peptide still poses some problems. While the cobaltocenium cation (with cobalt in the +III oxidation state) is very stable under almost all circumstances, ferrocene (with Fe(+II)) is readily oxidized to the ferrocenium cation (Fe(+III)), which decomposes much more readily than ferrocene. This oxidation is particularly facile in the presence of acids. For SPPS, this would demand the use of linkers which are *not* cleaved by acids. In fact, the first stable ferrocene-enkephalin conjugates in our group were prepared with the HMBA linker, which is cleaved by strong base such as conc. ammonia in methanol. However, such a strategy may not be always applicable when side chain protecting groups require acid cleavage, as do most common tyrosine protecting groups. Alternatively, acid cleavage can be used but protective additives have to be included. In the case of ferrocene, phenol can advantageously replace water in the cleavage mixture described above. It prevents the oxidation of ferrocene and hence the desired ferrocene-enkephalin can be obtained in good yield and purity without notable signs of disintegration. [24]

The peptide synthesis process that is explained above describes the most common protocol. Evidently, many variants are conceivable, for instance for the synthesis of (partly) protected peptides, combination of labeling groups, cyclic peptides, and many more. All reaction steps can be performed manually in syringes with a frit at the bottom, as described in the experiment below. Alternatively, a variety of computer-controlled peptide synthesizers are commercially available. Generally, these machines have a peptide synthesis protocol implemented which is optimized for the particular machine and this should be followed for peptide synthesis. In any case, we find it useful to transfer the peptide from the synthesizer to a syringe for manual coupling of new or sensitive metal complexes so as to have maximum control. After finalization of the peptide synthesis, addition of the metal complex, and cleavage, the peptide is precipitated by addition of very cold diethyl ether. The precipitated peptide is analyzed for purity by reversed-phase HPLC. In most cases, the chromatogram should show not much more than one single peak for the product. With luck the purity will be > 95% already and for most applications no further purification will be required. Purification may be performed by preparative HPLC. In most cases, MeOH/water mixtures or water/acetonitrile (with a little TFA added to both solvents) mixtures will be used as the mobile phase for peptides. The HPLC gradient can be adjusted according to the properties of the peptide to be isolated and the side products that need to be separated. For an overview chromatogram, a gradient from 5 to 95% acetonitrile is appropriate. For peptides, a standard reverse-phase C_{18} column will usually give good results. However, many other materials are available for specialized applications or product classes. Finally, mass spectrometry is used to confirm the mass (and thereby identity) of the obtained and purified products. If possible, an LC-MS combination can be used to obtain mass spectra directly along with the chromatogram. If this is not available, then isolated fractions from the HPLC can be injected into an electrospray-ionization MS (ESI-MS) or samples can be prepared for MALDI-MS. For the enkephalin-ferrocene peptides described herein, it may even be possible to obtain FAB mass spectra, although this is an exception in peptide chemistry.

Experiment

Objectives

✓ manual solid phase peptide sythesis (SPPS)
✓ incorporation of a metal complex into a manual SPPS procedure
✓ analysis of SPPS products by HPLC and mass spectrometry
✓ introduction to use of computer-controlled peptide synthesiser (optional)

Materials

All chemicals are needed in Peptide Synthesis Grade.

- dimethylformamide (DMF) – should be bought in "amine-free" quality, or purified to remove free dimethylamine
- dichloromethane (DCM)
- piperidine
- O-(1H-benzotriazol-1-yl)-1,1,3,3- tetramethyl-uronium tetrafluoroborate (TBTU) ($M = 321.1\,\text{g}\cdot\text{mol}^{-1}$)
- N-hydroxybenzotriazole (HOBt) ($M = 135.1\,\text{g}\cdot\text{mol}^{-1}$)
- diisopropylethylamine (DIPEA)
- trifluoroactetic acid (TFA)
- triisopropylsilane (TIS)
- phenol
- ferrocene carboxylic acid ($M = 230.2\,\text{g}\cdot\text{mol}^{-1}$)
- cobaltocenium carboxylic acid (not commercially available) ($M = 378.1\,\text{g}\cdot\text{mol}^{-1}$ as PF_6^- -salt)
- ruthenocene carboxylic acid (not commercially available) ($M = 275.3\,\text{g}\cdot\text{mol}^{-1}$)
- Fmoc-L-Leu-Wang Resin
- Fmoc-L-Phenylalanin-OH ($M = 387.4\,\text{g}\cdot\text{mol}^{-1}$)
- Fmoc-L-Tyrosin(*tert*-butyl)-OH ($M = 459.6\,\text{g}\cdot\text{mol}^{-1}$)
- Fmoc-Glycin-OH ($M = 297.3\,\text{g}\cdot\text{mol}^{-1}$)
- SPPS-reactor (5 ml Syringe with 25 ml Frit, for example from Multisyntech, Witten, Germany)
- vials (2 ml, single use), for the monomers and TBTU
- spatula s
- plastic single-use syringes (1 ml, ±0.01 ml, no rubber piston, for example Injekt-F from Braun, Melsungen, Germany),
- single-use tip-cut needles (0.9 × 79 mm, Neoject) for the syringes
- glass vials (5 ml) for DIPEA
- 2 glass vials (25 ml) for pre-washing and 20% piperidine in DMF, respectively
- 1 glass vial (40 ml) for main-washing
- 1 graduated cylinder for ether
- 2 centrifuge tubes (40 ml)
- 1 beaker or Erlenmeyer flask (250 ml) for liquid waste
- analytical balance, ±0.1 mg
- centrifuge

- vacuum pump, $p \leq 10$ mbar
- desiccator
- shaker

General Remarks

Of the metallocene carboxylic acids, only ferrocene carboxylic acid is commercially available. Ruthenium carboxylic acid and cobaltocenium carboxylic acid can be prepared according to the literature. [25–27] The synthesis of these metallocene carboxylic acids can be carried out before the SPPS, which would make the experiment more "inorganic", and synthetically more challenging. However, it should be noted that synthesis of the compounds will take several days to complete.

Swelling of the resin requires 1 h, in the meantime the monomers can be weighed and the solutions prepared. One synthesis cycle takes about 40 min, for 6 cycles up to 5 h are typically required. The final cleavage requires 90 min, and precipitation and isolation an additional 2 h. The spreadsheet given at the end of the chapter or a similar one should be used to monitor the progress of the synthesis and to avoid confusion (Table 6.1).

Special attention must be paid to avoid contamination of the stock solutions. Let each group use their own stock solutions! For the synthesis, the syringes with the reaction mixture can be placed on a motor-driven shaker. However, if no shaker is available, the synthesis works also well with only occasional shaking by hand. Shaking must be done gently; the resin beads should not be damaged.

Fmoc-Deprotection Solution

Piperidine is diluted with DMF to 20% (about 2 ml per synthesis step).

Preparation for the washing step

DMF is filled in two vials, one is used for the pre-wash (1 × 2 ml) and the other for the main wash (4 × 2 ml).

Preparation for the coupling step

In one vessel (single use reaction tube, 2 ml, e. g. Eppendorf tube), four equivalents of the monomer (protected amino acid or metallocene) needed for one step are weighed. In another vessel (single use reaction tube, 2 ml), four equivalents of TBTU and HOBt are weighed. The TBTU/HOBt mixture is dissolved in 1 ml of DMF and added to the monomer by syringe. For activation, 1 min prior to coupling, 100 μl of DIPEA is added to the monomer/TBTU/HOBt mixture. To minimize mishandling errors TBTU/HOBt should be weighed separately for every step. We also recommend that the syringe be disposed after a single use to avoid contamination.

Preparation of final cleavage solution

For metallocene-containing peptides use 90% TFA, 5% phenol, and 5% TIS (v/v/v). Ca. 2–3 ml per peptide should be used.

Preparation for precipitation

50 ml diethylether per peptide should be stored at −20 °C.

Resin preparation and swelling

A 5 ml plastic syringe with a polypropylene filter is used as the reactor, see Figure 6.7. The amount of resin used has to be calculated from the specified loading and filled into the reactor. Total amounts of chemicals needed have to be calculated and filled into the spreadsheet (Table 6.1). Using a tip-cut needle, about 2 ml of DMF and 1 ml of air are taken in the syringe. It is left for 1 h at room temperature (swelling of the resin) and the DMF is then pushed out of the syringe into the waste container.

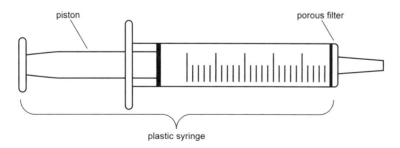

Fig. 6.7 Schematic drawing of the syringe reactor for manual solid phase peptide synthesis.

Fmoc deprotection and coupling of amino acids

Synthesis is organized in cycles as described here:

1. Fmoc deprotection. 2 ml of piperidine solution (20%) in DMF is taken into the syringe and the reaction is carried out for 10 min with shaking.

2. Washing. The deprotection solution is pushed out of the syringe into the waste container and the resin washed with 5×2 ml of DMF while shaking for about 1 min each time.

3. Coupling. To the solution of the monomer and TBTU/HOBt in DMF, 100 µl of DIPEA is added. The mixture is allowed to react for 1 min, then taken into the syringe and shaken for 20 min. For the metallocenes, longer coupling times are recommended (up to overnight, see text for optimization of conditions).

4. Washing. The coupling solution is pushed out of the syringe into the waste container and the resin washed with 5×2 ml DMF while shaking for about 1 min.

Steps 1 to 4 have to be repeated n times for n amino acids or metallocene-markers.

A scheme of the complete synthesis is depicted in Figure 6.8.

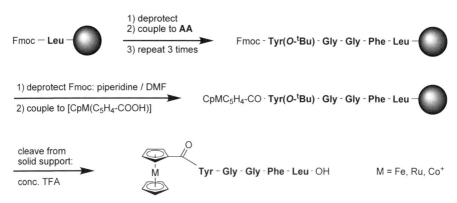

Fig. 6.8 Synthesis scheme for metallocene-Enkephalin conjugates.

Cleavage and deprotection

After the coupling of the metallocene, the resin is first washed with 5×2 ml DMF and then with 5×2 ml DCM. The syringes are placed in a vacuum desiccator and dried for 2 h. Then, 2 ml of the cleavage mixture are taken up in the syringe and the reaction vessel is left standing for 90 min at room temperature. After this period, the cleavage solution is poured into a weighed and labeled centrifuge tube and the resin washed with 1 ml TFA which is also added to the centrifuge tube.

Work up

The peptide is precipitated by addition of 20 ml diethylether to the combined TFA solutions, the suspension is centrifuged and the supernatant decanted. The residue is washed with diethylether (2×10 ml) by slurrying, centrifugating, and decanting. The product is dried in vacuo and weighed.

Analysis of the synthesis product

The usual method to determine purity and success of the synthesis is to perform a reverse phase chromatography (HPLC) and mass spectrometry. The amount needed depends on the sensitivity of the system, but a few μl of 1 mg of the metallocene-bioconjugate in 1 ml of Methanol/H_2O or acetonitrile/H_2O should be detectable with most systems. FAB, ESI, and MALDI mass spectrometers are suitable. It should be possible to detect the singly charged molecule signal with all methods. In the HPL chromatogram, the purity can be assessed by integration of the detected peaks. If the retention time of the compound on the system is known, it can also be used for identification. We use reversed-phase C_{18} columns (5 μm material, 4 mm diameter for analytical HPLC, flow 1 ml/min). A linear gradient from 5 to 95% acetonitrile/water over 20 min is used, 0.1% TFA is added to both solvents. With this system, we find retention times of 13.3 min for ferrocenoyl enkephalin, 9.6 min for the cobaltocenium enkephalin described herein, 10.5 min for acetylated enkephalin and 9.3 min for unsubstituted enkephalin alone.

Variations of the experiment

The procedure described above can readily be adapted to any other desired peptide. The main difficulty with most other peptides will be functional side chains, which require protecting groups and deprotection. Procedures for establishing reliable methods are described. There is no principal limitation in the choice of peptides.

Also, other metal complexes may be used. The experiment described herein is optimized for attachment of carboxylic acid derivatives. However, other chemistry can also be used. For example, we have successfully employed the Cu-catalyzed [3+2] dipolar cycloaddition reaction ("Click chemistry") for the synthesis of metal bioconjugates. [28]

In the above experiment, the metal complex is attached on the *N*-terminal end of the peptide. However, this is not the only possibility. By using lysine amino acids with very acid-labile side chains (such as the Mmt group), this lysine can be selectively deprotected and a metal complex carboxylic acid attached to this lysine side chain.

Spreadsheet

A spreadsheet will help keep track of the progress of the synthesis. An empty sample spreadsheet is given in Table 6.1. Positions marked with an "x" are not applicable for the corresponding step. An example of how the spreadsheet should be filled in is given in Table 6.2. The addition of DIPEA to the coupling mixture is marked with a "+". Each single washing is marked with a bar.

Table 6.1 Empty spreadsheet for [Leu^5]-Enkephalin.

		resin loading: … mmol/g; 100 mg resin $=$ … mmol, 4 eq $=$ … mmol							
Nr.	Monomer	M_r	m (4 eq)	DIPEA	coupling	washing	piperidine	washing
1	Leu-Wang	x	x	x	x	x		
2	Fmoc-Phe-OH							
3	Fmoc-Gly-OH							
4	Fmoc-Gly-OH							
5	Fmoc-Tyr(tBu)-OH							
6	Metallocene-OH						x	x
–	TBTU (for each step)							
–	HOBt (for each step)							

Table 6.2 Partly filled-in spreadsheet (example).

Fmoc-Leu-Wang resin: $0.68\,\mathrm{mmol/g}$; $100\,\mathrm{mg} = 0.068\,\mathrm{mmol/g}$, $4\,\mathrm{eq} = 0.272\,\mathrm{mmol}$ [a]

Nr.	Monomer	M/gmol^{-1}	m/mg [b]	DIPEA	couple [c]	wash [c]	deprotect	Wash
1	Leu-Wang	x	x	x	x	x	9^{18}	///
2	Fmoc-Phe-OH	387.4	105.4	+	9^{32}	///	9^{56}	///
3	Fmoc-Gly-OH	297.3	80.9	+	10^{11}	///	10^{37}	
4	Fmoc-Gly-OH	297.3	80.9					
5	Fmoc-Tyr(tBu)-OH	459.6	127.4					
6	Metallocene-OH	*	*				x	x
–	TBTU (for each step)	321.1	87.3					
–	HOBt (for each step)	135.1	36.7					

[a] The loading is specified by the manufacturer
[b] 4 equivalents of the monomer or TBTU are taken
[c] Time the step was performed
* Metallocene (free acid):
  Cobaltocenium carboxylic acid hexafluorophosphat (M = 378.1 g/mol): 102.8 mg
  Ferrocene carboxylic acid (M = 230.0 g/mol): 62.6 mg
  Ruthenocene carboxylic acid (M = 275.3 g/mol): 74.87 mg

Additional questions

✓ Explain the advantages of solid phase synthesis as compared to synthesis in solution. How long would the syntheses and purifications take if all steps were carried out sequentially in solution?

✓ In this experiment an Fmoc solid phase synthesis strategy is applied. Merrifield, who invented the chemical synthesis on solid support, initially developed a strategy based on the N-terminal Boc protection group. Describe shortly the Boc strategy and the advantages and disadvantages of both methods.

✓ Organometallics derived from ferrocene (A) and cobaltocenium hexafluorophosphate (B) are used as markers. Explain the elelectronic stucture of the parent compounds (A) and (B), using the 18-electron rule.

✓ The molar mass of the final product is used for calculation of the yield, but the exact mass is required to assign the mass spectra. Explain the difference between the molar mass and the exact mass. Why does the difference between the two numbers become more significant in larger (bio)molecules?

✓ In this experiment, we use three organometallic labels which are well-suited for electrochemical detection of the conjugates. On the other hand, dyes like rhodamine or fluorescein are frequently used for fluorescence detection. Describe the physical principles of both methods and give examples that employ such techniques.

Bibliography

[1] G. Jaouen (ed.), *Bioorganometallics*, Wiley-VCH, Weinheim, **2006**.

[2] N. Metzler-Nolte in *Comprehensive Organometallic Chemistry III*, G. Parkin (ed.), Elsevier, Amsterdam, **2006**; Vol. 1, p 883–920.

[3] A. Vessières, M. Salmain, P. Brossier, G. Jaouen, Carbonyl metallo immuno assay: A new application for fourier transform infrared spectroscopy, *J. Pharm. Biomed. Anal.* **1999**, 21, 625–633.

[4] N. Metzler-Nolte, Labeling of biomolecules for medicinal applications – Bioorganometallic chemistry at its best, *Angew. Chem. Int. Ed.* **2001**, 40, 1040–1044.

[5] M. Salmain, N. Fischer-Durand, L. Cavalier, B. Rudolf, J. Zakrzewski, G. Jaouen, Transition metal-carbonyl labeling of biotin and avidin for use in solid-phase carbonyl metallo immunoassay (CMIA), *Bioconjugate Chem.* **2002**, 13, 693–698.

[6] S. I. Kirin, I. Ott, R. Gust, W. Mier, T. Weyhermüller, N. Metzler-Nolte, Cellular uptake quantification of metallated peptide and peptide nucleic acid (PNA) bioconjugates by atomic absorption spectroscopy, *Angew. Chem. Int. Ed.* **2008**, 47, 955–959.

[7] J. Niesel, A. Pinto, H. W. Peindy N'Dongo, K. Merz, I. Ott, R. Gust, U. Schatzschneider, Photoinduced CO release, cellular uptake and cytotoxicity of a tris(pyrazolyl)methane (tpm) manganese tricarbonyl complex, *Chem. Commun.* **2008**, 1798–1800.

[8] S. I. Kirin, F. Noor, N. Metzler-Nolte, W. Mier, Manual solid phase peptide synthesis, *J. Chem. Educ.* **2007**, 84, 108–111.

[9] N. Metzler-Nolte, Medicinal applications of organometal-peptide bioconjugates, *Chimia* **2007**, 61, 736–741.

[10] J. Chantson, M. V. Varga Falzacappa, S. Crovella, N. Metzler-Nolte, Solid phase synthesis, characterisation and anti-bacterial activity of metallocene-peptide bioconjugates, *ChemMedChem* **2006**, 1, 1268–1274.

[11] M. Neukamm, A. Pinto, N. Metzler-Nolte, Synthesis and cytotoxicity of a cobaltcarbonyl-alkyne enkephalin bioconjugate, *Chem. Commun.* **2008**, 232–234.

[12] F. Noor, R. Kinscherf, G. Bonaterra, N. Metzler-Nolte, Enhanced cellular uptake of organometallic bioconjugates of the NLS peptide in HepG2 cells, *ChemBioChem* **2008**, 10, 493–502.

[13] F. Noor, A. Wüstholz, R. Kinscherf, N. Metzler-Nolte, A cobaltocenium peptide bioconjugate shows enhanced cellular uptake and directed nuclear delivery, *Angew. Chem. Int. Ed.* **2005**, 44, 2429–2432.

[14] R. B. Merrifield, Solid phase peptide synthesis. I. The synthesis of a tetrapeptide, *J. Am. Chem. Soc.* **1963**, 85, 2149–2154.

[15] G. Lu, S. Mojsov, J. P. Tam, R. B. Merrifield, Improved synthesis of 4-alkoxybenzyl alcohol resin, *J. Org. Chem.* **1981**, 46, 3433–3436.

[16] S.-S. Wang, P-alkoxybenzyl alcohol resin and p-alkoxybenzyloxycarbonylhydrazide resin for solid phase synthesis of protected peptide fragments, *J. Am. Chem. Soc.* **1973**, 95, 1328–1333.

[17] C. A. G. N. Montalbetti, V. Falque, Amide bond formation and peptide coupling, *Tetrahedron* **2005**, 61, 10827–10852.

[18] E. Valeur, M. Bradley, Amide bond formation: Beyond the myth of coupling reagents, *Chem. Soc. Rev.* **2009**, 38, 606–631.

[19] A. T. R. Knorr, W. Bannwarth, D. Gillessen, New coupling reagents in peptide chemistry, *Tetrahedron Lett.* **1989**, 30, 1927–1930.

[20] L. A. Carpino, 1-hydroxy-7-azabenzotriazole. An efficient peptide coupling additive, *J. Am. Chem. Soc.* **1993**, 115, 4397–4398.

[21] C. D. Chang, J. Meienhofer, Solid-phase peptide synthesis using mild base cleavage of n alpha-fluorenylmethyloxycarbonylamino acids, exemplified by a synthesis of dihydrosomatostatin., *Int. J. Pept. Protein Res.* **1978**, 11, 246–249.

[22] T. W. Green, P. G. M. Wuts, *Protective Groups in Organic Synthesis*, Wiley, Hoboken, New Jersey, **2007**.

[23] J. Hughes, T. Smith, H. Kosterlitz, L. Fothergill, B. Morgan, H. Morris, Identification of two related pentapeptides from the brain with potent opiate agonist activity, *Nature* **1975**, 258, 577–580.

[24] A. Pinto, N. Metzler-Nolte, Modification with organometallic compounds facilitates crossing of the blood-brain barrier of [Leu⁵]-enkephalin derivatives, *ChemBioChem* **2009**, in press.

[25] P. C. Reeves, Carboxylation of aromatic compounds: Ferrocene carboxylic acids, *Org. Synth.* **1977**, 56, 28.

[26] M. D. Rausch, E. O. Fischer, H. Grubert, The aromatic reactivity of ferrocene, ruthenocene, and osmocene, *J. Am. Chem. Soc.* **1960**, 82, 76–82.

[27] J. E. Sheats, M. D. Rausch, Synthesis and properties of cobalticinium salts. I. Synthesis of monosubstituted cobalticinium salts, *J. Org. Chem.* **1970**, 35, 3245–3249.

[28] S. D. Köster, J. Dittrich, G. Gasser, N. Hüsken, I. C. H. Castañeda, J. L. Jios, C. O. D. Védova, N. Metzler-Nolte, Spectroscopic and electrochemical studies of ferrocenyl triazole amino acid and peptide bioconjugates synthesized by click chemistry, *Organometallics* **2008**, 27, 6326–6332.

7 Preparation of metal-protein bioconjugates

Summary. This chapter will familiarize you with organometallic protein bioconjugates. You will be introduced to the principles of protein structure and questions of reactivity and selectivity during the derivatization of proteins in solution. We use the enzyme lysozyme from hen egg white as a model protein and metallocene carboxylic acids as metal compounds. In the experimental part you will attach the activated metal complexes to lysozyme and characterize the products by HPLC and mass spectrometry.

Learning targets

✓ Chemical structure of proteins
✓ Chemical synthesis of protein conjugates in solution
✓ Chromatographic analysis and purification of proteins and their metal conjugates
✓ Spectroscopic characterization of metal-protein conjugates (see also Chapter 8)

Background

In the previous chapter, we examined the preparation of metal conjugates of peptides. The peptides were prepared completely synthetically, and thus we were free to introduce the metal complex at whatever position we wished, of course depending on the chemistry and supposing that technical problems can be overcome. In this experiment, we aim to link metal complexes to a naturally occurring protein. The problem we face in this experiment is selectivity: a protein may consist of a few hundred to many thousand amino acids. Given that there are only 20 naturally occurring amino acids, each amino acid will statistically be present several to many times – depending on the size of the protein. Unlike in chemical synthesis, where we request absolute control, here we have to deal with a degree of uncertainty, and we have to consider whether and how this influences our experiment and the expected results.

First, let us consider how proteins are made in the cell, and which options we may have to influence their composition. Proteins are synthesized from single amino acids on the ribosome by a complicated, multi-enzyme machinery. The amino acid sequence is encoded in the genome of the organism in the form of its DNA (see Chapter 4). The DNA is transcribed to a corresponding messenger or mRNA sequence which in turn is read by the ribozyme when the organism needs a certain protein to be made. Each amino acid is delivered to the ribosome by a special RNA, called the transfer or tRNA. When the tRNA matches with the mRNA sequence required for the next amino acid, then the tRNA is bound and the amino acid is added to the growing protein chain. The protein is built on the ribozyme from the N- to the C-terminus. This is the reverse order from chemical peptide synthesis as shown in the previous Chapter 6. In solid phase peptide synthesis, the *incoming* amino acid is activated and added to the N-terminus of the peptide with formation of the new peptide bond. On the ribozyme, the last amino acid of the protein chain is activated by reaction with the ribozyme, and the next amino acid is offered by the tRNA in such a way, that only the amino group can react. Unlike SPPS, no further protecting groups (e. g. for amino side chains in lysines) are required, because the enzyme will hold the activated and incoming amino acid in place such that no side reactions are possible.

The whole process in its astonishing details is described in biochemistry text books. [1] Possibilities of interference with the DNA are discussed in Chapters 4 and 5 of this book. Here, let us consider the possibilities to influence the amino acid sequence. Each single amino acid is encoded by a combination of three bases in the RNA (or originally the DNA), called a triplet. These DNA triplets are unique in the sense that one combination of bases will always translate into the same amino acid. Some amino acids, however, may be coded for by several different combinations of bases. Obviously we can change the DNA, and consequently the mRNA sequence, so that complete amino acids are either deleted, added, or replaced by other amino acids. This "cloning" is a routine operation in biochemistry and is frequently applied to obtain modified proteins. However simple this sounds in principle, it may not always work reliably, and it is certainly time consuming. Of course, the successfully modified protein may also have significantly different biophysical properties, such as decreased solubility or altered kinetic parameters in enzyme catalysis.

Since there are four nucleobases, $4^3 = 64$ combinations are possible in theory. Because only 20 amino acids need to be coded for, there is clearly some redundancy. So, one might ask whether it is possible to encode new, unnatural amino acids, by using a combination of RNA nucleobases that does not yet define an existing amino acid. Indeed, such techniques have been developed and it is possible to introduce completely unnatural amino acids into proteins. [2–4] However, a whole range of "new" enzymes must be produced, including the tRNA for the new triplet. This process is extremely cumbersome, and the whole technology is well-cstablished in only a few labs worldwide. Moreover, it is *very* time-consuming, with uncertain success, and therefore certainly only initiated with very good reason. As far as we know, no metal-containing amino acid has been used in this process so far.

So, a better option would be to modify existing amino acids in a protein with metal complexes. [5] To this end, a few functional groups spring to mind. Most notably, cysteine thiol groups offer the possibility for functionalization with thiol-specific reactive compounds. This indeed has been realized and Figure 7.1 shows a few thiol-specific ferrocene-containing reagents. [6, 7] One advantage of thiol-specific reagents is that cysteines are relatively rare in most proteins. On the other hand, they are often engaged in the formation of sulfur-sulfur bonds, i.e. cystines.

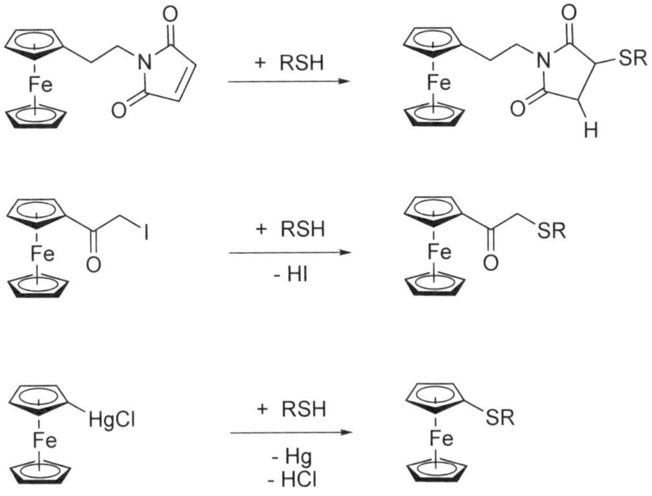

Fig. 7.1 Some ferrocene derivatives with thiol-specific reactive groups that are used to modify cysteine residues in proteins.

Furthermore, and for this reason, they are not normally exposed to the surface of the protein, but rather buried in the inside, and thus may not be accessible to a thiol-specific reagent without denaturing of the protein.

Another option is to use the ε-amino groups of lysines for modification. First of all, there are generally several lysines present in a protein. Moreover, the lysine residues are usually exposed to the protein surface and because they are protonated at neutral pH, they help to solubilize the protein in water. These amino groups can be reacted with activated esters, as in this experiment, to give amide-modified lysine residues. One important consideration follows from this fact, i.e. that the pH value must be carefully controlled in order to ensure sufficient nucleophilicity of the lysine amino groups. Looking at this problem in a more general way, we need to avoid any reagents or conditions which might cause side reactions with the activated ester that we use for modifying the protein. Although this sounds almost trivial, it is often overlooked that buffers, stabilizers, additives, etc. may contain chemicals which will interfere with the planned reaction. Even worse, some commercial suppliers do not list all chemicals that were added, for example in special biochemical assays. On the other hand, it may be advantageous to develop special reagents that are robust against varying conditions. For example, instead of the activated ester which is used in this experiment, the use of pyrylium salts has been proposed for the modification of lysine amino groups in enzymes. [8, 9] These pyrylium salts react with primary amino groups to yield pyrdinium salts, and this reaction is rather insensitive to the pH value of the reaction. Unlike the activated ester reaction, it can also be carried out at neutral pH, which is an advantage if the protein of choice is either insoluble or denatures under basic conditions.

In this experiment, we use again metallocenes and an activated ester for reaction with amino groups much like those described in the previous chapter for the labeling of peptides with organometallic compounds. As a model protein, we choose the enzyme white hen egg lysozyme (HEWL). Lysozmes, also known as muramidases or N-acetylmuramide glycanhydrolases are a family of enzymes (EC number 3.2.1.17) which damage bacterial cell walls by hydrolysis of particular glycosidic bonds in peptidoglycans. Lysozyme is part of the innate immune system, and it protects against bacterial infections by hydrolyzing the bacterial cell walls especially of Gram-positive bacteria, which consist of peptidoglycans. It is present in secretions of humans, e. g. tears and saliva. It is also present, and more readily obtained, from egg white. HEWL in particular is inexpensive and commercially available in > 95% purity. It is a relatively small protein with 129 amino acids in a single peptide chain. It can be easily crystallized under a variety of conditions and has only a limited number of reactive side chains. [10] Its molecular weight is 14314 Da, and it has six lysine side chains (Lys1, 13, 33, 96, 97, 116), plus one N-terminal amino group. [11]

Most of these amino groups are accessible for reaction with activated esters. A schematic reaction scheme is shown in Figure 7.2. If the protein is derivatized with more than about two lipophilic groups, such as the ferrocenoyl groups used in this experiment, then its solubility will decrease and precipitation occurs. On the other hand, if only a 1 : 1 ratio is used for the derivatization reaction, the modified product will contain several isomers. We can employ trypsin digestion coupled to mass spectrometry to determine which lysine residues have been modified, as described in detail in the literature. [9] The method relies on the fact that the modified lysine will have a weight of 243 mass units higher than the unmodified lysine, and thus any peptide fragment containing such a ferrocenoyl lysine will be shifted in its weight accordingly. Without having to assign every single trypsin fragment in the mass spectrum, it is thus possible to discern the main sites of derivatization.

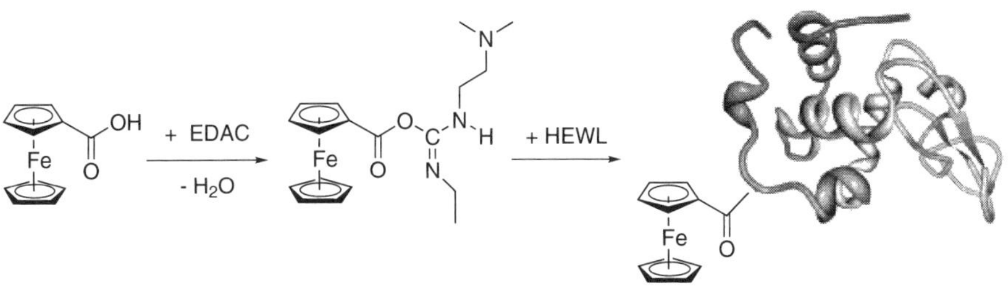

Fig. 7.2 Schematic reaction scheme for the modification of hen egg white lysozyme (HEWL) with ferrocene carboxylic acid, which is activated by EDAC. Reaction of the activated ferrocene carboxylate with a lysine group of the protein is assumed, and only one of the possible 1 : 1 adducts is shown.

Experiment

Objectives

✓ covalent modification of white hen egg lysozyme with a metallocene
✓ analysis of the products by HPLC, SDS-PAGE and mass spectrometry
✓ trypsin digestion and determination of the sites of modification

Materials

– acetic acid
– 2-($1H$-7-azabenzotriazole-1-yl)-1,1,3,3-tetramethyluronium hexafluorophosphate methanaminium (HATU, $M = 380.2\,\mathrm{g \cdot mol^{-1}}$)
– acrylamide, 40%
– ammonium bicarbonate, NH_4HCO_3 ($M = 79.06\,\mathrm{g \cdot mol^{-1}}$)
– ammonium persulfate (APS), $(NH_4)_2SO_5$ ($M = 148.14\,\mathrm{g \cdot mol^{-1}}$)
– bromophenol blue
– cobaltocenium carboxylic acid (not commercially available, see Chapter 6), ($M = 378.1\,\mathrm{g \cdot mol^{-1}}$ as PF_6^--salt)
– diisopropylethylamine (DIPEA)
– dithiothreitol (DTT)
– 1-(-Dimethylaminopropyl)-3-ethylcarbodiimide hydrochlorid (EDAC, $M = 191.7\,\mathrm{g \cdot mol^{-1}}$)
– ethanol
– ethylenediaminetetraacetic acid (EDTA)
– ferrocene carboxylic acid ($M = 230.04\,\mathrm{g \cdot mol^{-1}}$)
– formic acid
– hen egg white lysozyme (HEWL)
– 2,6-lutidine ($M = 107.16\,\mathrm{g \cdot mol^{-1}}$)
– 3-(N-morpholino)propane sulfonic acid (MOPS)
– potassium carbonate, K_2CO_3 ($M = 138.21\,\mathrm{g \cdot mol^{-1}}$)
– potassium bicarbonate, $KHCO_3$ ($M = 100.12\,\mathrm{g \cdot mol^{-1}}$)
– potassium ferrocyanide $K_4[Fe(CN)_6]$ ($M = 368.34\,\mathrm{g \cdot mol^{-1}}$)
– ruthenocene carboxylic acid ($M = 275.3\,\mathrm{g \cdot mol^{-1}}$)
– silver nitrate, $AgNO_3$ ($M = 169.87\,\mathrm{g \cdot mol^{-1}}$)

- sodium acetate, $NaCH_3CO_2$ ($M = 82.03\,\text{g}\cdot\text{mol}^{-1}$)
- sodium carbonate, Na_2CO_3 ($M = 105.99\,\text{g}\cdot\text{mol}^{-1}$)
- sodium thiosulfate, $Na_2S_2O_3$ ($M = 158.11\,\text{g}\cdot\text{mol}^{-1}$)
- sodium dodecyl sulfate (SDS)
- triisopropylsilane (TIS)
- tetramethylethylenediamine (TEMED)
- trifluoroacetic acid (TFA)
- tris(hydroxymethyl)aminomethane (TRIS)
- analytical balances, $\Delta m = \pm 0.1\,\text{mg}$
- 1 beaker or Erlenmeyer flask (250 ml) for liquid waste
- 1 centrifuge
- 2 centrifuge tubes (40 ml)
- glass vials (5 ml) for DIPEA
- plastic single-use syringes (1 ml, ± 0.01 ml, no rubber piston, for example Injekt-F, Braun, Melsungen, Germany)
- power supply, glass plates, comb, and stand for SDS-PAGE
- single-use tip-cut needles (0.9×79 mm, Neoject) for the syringes
- shaker
- spatula
- speedvac
- vials (2 ml, single use), for the monomers and HATU
- 2 glass vials (25 ml) for pre-washing and 20 % piperidine in DMF, respectively
- 1 glass vial (40 ml) for main-washing
- 1 vacuum pump, $p \leq 10$ mbar
- plastic tubes (50 ml, 15 ml, 1.5 ml)

Buffers and solutions

Bleaching solution

1 : 1 mixture of:

30 mM sodium thiosulfate $Na_2S_2O_3$
100 mM potassium ferrocyanide

Carbonate buffer, pH 10.1

0.05 M potassium carbonate

0.05 M potassium bicarbonate

Buffer I, pH 6.8

0.5 M TRIS, adjusted to pH = 6.8 with hydrochloric acid

Buffer II, pH 8.8

1.5 M TRIS, adjusted to pH = 8.8 with hydrochloric acid

10× MOPS running buffer, pH 6.5

0.5 M MOPS

0.5 M TRIS

0.035 M SDS

0.01 M EDTA

Sample buffer (5×), pH 6.8

160 mM TRIS

5% (w/v) SDS

0.5 M DTT

0.05% bromophenol blue

adjusted to pH = 6.8 with hydrochloric acid

General remarks

Ferrocene carboxylic acid is commercially available. Ruthenium carboxylic acid and Cobaltocenium carboxylic acid can be prepared according to the literature. [12, 13] The synthesis of these metallocene carboxylic acids may be carried out before the protein labeling, or they can be used from the previous experiment (Chapter 6) in this book.

Activation of ferrocene carboxylic acid with EDAC

Ferrocene carboxylic acid (46 mg, 0.2 mmol) and EDAC (38 mg, 0.2 mmol) are mixed in a 1 : 1 ratio and stirred in 2 ml carbonate buffer (pH 10.1) for

30 min at room temperature in the dark. The resulting solution is used as is.

Activation of ferrocene carboxylic acid with HATU

Ferrocene carboxylic acid (78 mg, 0.34 mmol) and HATU (129 mg, 0.34 mmol) are mixed in a 1 : 1 ratio, dissolved in 1 ml lutidine: DIPEA:DMF (1 : 1 : 1, v/v/v) and stirred for 30 min at room temperature in the dark. The resulting solution is used as is.

Derivatization of the protein

The protein hen egg white lysozyme (HEWL) (0.02 mmol, 294 mg) is dissolved in 0.2 ml carbonate buffer (pH 10.1). For stoichiometric reactions in a 1 : 1 ratio, about 10% of the solution containing EDAC-activated ferrocene carboxylic acid is added to the white slurry of HEWL, the combined solutions are stirred for 17 to 22 h in the dark. Alternatively, ca. 60 µl of the HATU-activated ferrocene carboxylic acid solution is added to the dissolved protein and stirred for 17 to 22 h in the dark.

Work-up

The solution is centrifuged for 10 min at 4000 rpm. The supernatant is then filtered through a 0.2 µm pore filter under mild pressure.

Analysis of the results

HPLC and mass spectrometry

HPLC is performed for analysis of the products. For analytical runs, we use a C_{18} reversed-phase column with 250×40 mm (5 µm material). Measurements are carried out with a flowrate of 1.0 ml/min. Chromatogramms are recorded at 220 nm and 254 nm. Mixtures of ultrapure water and acetonitrile with 0.1% TFA (v/v) are used as eluents. The chosen gradient is linear 5 to 95% acetonitrile, over a run time of 30 min. After that, equilibrium is restored at 5% acetonitrile again. For comparison, chromatogramms of pure dissolved HEWL and Fc-COOH should be recorded, too.

Table 7.1 Composition of acrylamide solutions for SDS-PAGE.

	4% acrylamide, 10 ml	12% acrylamide, 50 ml
Acrylamide, 40%	1.0 ml	15 ml
Buffer I / II	3 ml [a]	14.6 ml [b]
H_2O, millipore	5.9 ml	20.1 ml
APS, 20% (w/v)	100 μl	300 μl
TEMED	10 μl	30 μl

[a] of buffer I
[b] of buffer II

Typical retention times on our system using the above conditions are:

HEWL: 13 min;

Fc-COOH: 15 min;

Ferrocenoyl-HEWL bioconjugates: > 15 min;

Individual fractions can be collected either by hand or (if available) by an automated fraction collector. The fractions can be analyzed by mass spectrometry. For electrospray ionization mass spectrometry (ESI-MS), the solution from the HPLC can usually directly be injected. Alternatively, it can be used for MALDI-MS (matrix-assisted laser desorption ionization mass spectrometry), usually after evaporation and re-dissolving the residue in a suitable solvent for preparation of the MALDI-MS targets. Alternatively, a combined LC-MS instrument may be available and the mass for all peaks can be obtained directly. Typical MS data (m/z): HEWL: $[M+H]^+ = 14\,701$; Ferrocenoyl-HEWL bioconjugates: $[M+Fc+H]^+ = 14\,913$; In general: $[M+Fc_n+H]^{m+} = (14\,700 + (212) \times n + 1)/m$

SDS-PAGE gel electrophoresis

The upper part of the gel contains 4% acrylamide to collect the sample in the gel prior to the separation. Table 7.1 sums up the preparation conditions. The procedure for preparing gel electrophoresis is also described in Chapter 5. APS and TEMED are added last, directly before the solution is filled into the chamber of the apparatus.

Once the acrylamide solution is polymerized completely, the gel is covered with MOPS as the running buffer and the comb is removed.

Sample preparation:

The filtrated supernatant is diluted in carbonate buffer (pH 10.1) to 1 : 20. The dilution is then diluted again to 1 : 100. To 10 μl of each dilution are added 2.5 μl of sample buffer. The 12.5 μl samples are applied into the wells of the gel. 2.5 μl of protein ladder are added to the first and last well. The electrophoresis is run at 70 V in 4% acrylamide concentration and at 100 V in the 12% acrylamide gel.

Silver staining of the SDS-gel after electrophoresis

All solutions (400 ml) are prepared freshly in ultrapure water (Table 7.2).

Proteolytic digestion of proteins after silver staining

The protein bands are cut manually, transferred into 1.5 ml tubes and incubated in bleaching solution (15 μl) for 1 min. The gel pieces are incubated alternately for 10 min in ammonium bicarbonate (10 mM, pH 8.6) and then a mixture of ammonium bicarbonate (10 mM, pH 8.6) and acetonitrile (1 : 1) until the gel pieces are completely bleached. The pieces are centrifuged to dryness under vacuum (Speedvac) for at least 10 min.

Table 7.2 Silver staining protocol.

fixation	50% (v/v) ethanol 10% (v/v) acetic acid	overnight
incubation	0.5 M sodium acetate 0.1% (w/v) sodium thiosulfate 30% (v/v) ethanol	1 h wash 3× for 10 min with ultrapure water
staining	0.1% (w/v) sodium nitrate	20 min wash with MilliporeR water for [???] s
rinsing	2.5% (w/v) sodium carbonate	1 min
development	2.5% (w/v) sodium carbonate 0.01% formaldehyde	5 to 20 min
stop	0.05 M EDTA	at least 20 min wash 10 min with ultrapure water

Tryptic Digestion

The gel pieces are soaked in $2\,\mu l$ of a trypsin solution ($0.03\,\mu g/\mu l$ in $10\,mM$ ammonium bicarbonate, pH 8.6) overnight at $37\,°C$.

Extraction of peptides from gel

The gel pieces are sonicated in a mixture of 5% (v/v) formic acid and 50% (v/v) acetonitrile for 15 min. The supernatants are retained and the procedure is repeated. The supernatants are then combined and brought to $15\,\mu l$ in the speedvac and analyzed by mass spectrometry. The found m/z values are compared to predicted data from tryptic digested lysozyme: http://prospector.ucsf.edu/cgi-bin/msform.cgi?form=msdigest (choose parameters: Database: UniProtKB.200806.10, Digest: Trypsin, Max. missed cleavages: 0, List of Entries: P00698 (lysozyme precursor)).

Variations of the experiment

As described above, the pH value plays a crucial role for reaction of activated acids with lysine amino groups. You can determine the optimum pH value for this reaction by yourself by running the above reaction at different pH values and determine the amount of modified lysozyme, either by SDS-PAGE (gel electrophoresis) and quantification of the bands, or by mass spectrometry. Although this is not a quantitative method by its very nature, the relative peak intensities of the same species across the labeling experiments at different pH values will give a rather good estimate of the coupling efficiency, given that the work-up and MS conditions are the same for all samples.

Many more and other metal complexes can be used in the experiment. For example, the metallocene carboxylic acid derivatives from the previous experiment may equally be used in this experiment. As in the peptide labeling experiment, the activation time as well as the reaction time should be extended when the cobaltocenium carboxylic acid is used. On the other hand, solubility of the modified enzyme is less an issue with this complex as cobaltocenium carries a positive charge by itself. Pyrylium salts can also be used as described in the literature. [9]

You may want to use some of the ferrocene-modified protein for the electrochemical experiments described in the next chapter, so save it!

Additional questions

✓ Describe the biological role of lysozyme in detail giving the chemical formulae of the peptidoglycans and how they are cleaved. What is the exact mechanism of cleavage?

✓ Consult a crystal structure of HEWL and identify the main sites of modification with the ferrocenoyl group as described herein and determined in your experiment. Where are they located? Can you suggest reasons why mostly those amino groups were attacked? Where is the N-terminal amino group located? Is it also likely to react?

Bibliography

[1] L. Stryer, *Biochemistry*, 6th ed., W. H. Freeman, New York, **2007**.

[2] J. A. Ellman, D. Mendel, S. Anthony-Cahill, C. J. Noren, P. G. Schultz, Biosynthetic methods for introducing unnatural amino acids site-specifically into proteins, *Methods Enzymol.* **1991**, 202, 301–336.

[3] V. W. Cornish, M. I. Kaplan, D. L. Veenstra, P. A. Kollman, P. G. Schultz, Stabilizing and destabilizing effects of placing β-branched amino acids in protein α-helices, *Biochemistry* **1994**, 33, 12022–12031.

[4] J. Xie, L. Wang, N. Wu, A. Brock, G. Spraggon, P. G. Schultz, The site-specific incorporation of *p*-iodo-l-phenylalanine into proteins for structure determination, *Nature Biotechnol.* **2004**, 22, 1297–1301.

[5] J. M. Antos, M. B. Francis, Transition metal catalyzed methods for site-selective protein modification, *Curr. Opin. Chem. Biol.* **2006**, 10, 253–262.

[6] D. R. van Staveren, N. Metzler-Nolte, Bioorganometallic chemistry of ferrocene, *Chem. Rev.* **2004**, 104, 5931–5985.

[7] M. Salmain, N. Metzler-Nolte In *Ferrocenes*; P. Stepnicka (ed.), Wiley, Chichester, **2008**, p. 499–639.

[8] M. Salmain, K. L. Malisza, S. Top, G. Jaouen, M.-C. Sénéchal-Tocquer, D. Sénéchal, B. Caro, [η^5-cyclopentadienyl]metal tricarbonyl pyrylium salts: Novel reagents for the specific conjugation of proteins with transition organometallic labels, *Bioconjugate Chem.* **1994**, 5, 655–659.

[9] M. Salmain, B. Caro, F. Le Guen-Robin, J. C. Blais, G. Jaouen, Solution- and crystal-phase covalent modification of lysozyme by a purpose-designed organoruthenium complex. A maldi-tof ms study of its metal binding sites, *ChemBioChem* **2004**, 5, 99–109.

[10] M. M. Ries-Kautt, A. F. Ducruix, Relative effectiveness of various ions on the solubility and crystal growth of lysozyme, *J. Biol. Chem.* **1989**, 264, 745–748.

[11] http://lysozyme.co.uk/

[12] J. E. Sheats, M. D. Rausch, Synthesis and properties of cobalticinium salts. I. Synthesis of monosubstituted cobalticinium salts, *J. Org. Chem.* **1970**, 35, 3245–3249.

[13] M. D. Rausch, E. O. Fischer, H. Grubert, The aromatic reactivity of ferrocene, ruthenocene, and osmocene, *J. Am. Chem. Soc.* **1960**, 82, 76–82.

8 Electrochemical investigation of metal complexes

Summary. Rich redox chemistry and electron transfer processes are some of the most prominent features of metal complexes. This chapter will familiarize you with electrochemical investigations on metal complexes. In the experimental part, redox properties of several metal complexes, as well as their dependence on experimental parameters will be compared. By using metal-peptide and metal-protein conjugates from the previous experiments, diffusion coefficients can be determined.

Learning targets

✓ Electron transfer in nature and in metallo-enzymes in particular
✓ Electronic structure of metallocenes
✓ Redox chemistry of metallocenes
✓ Electrochemical characterization of metal bioconjugates (see also Chapters 6 and 7 for suitably modified peptides and enzymes)

Background

Metal ions fulfill numerous important functions in biological systems. Arguably the most prominent function of the transition metals is in redox reactions and electron transfer. Metal complexes, especially of the transition elements, have in most cases several redox states available to them. For manganese, for example, complexes are known for essentially all oxidation states between zero (metallic Mn) and +VII (e. g. in permanganate, MnO_4^-). Not all of these oxidation states are used in biological systems, however. Again for Mn, only the three oxidation states, +II, +III, and +IV, occur frequently in biological systems. The same is true for iron, which uses only +II and +III in all biologically relevant systems (with the possibility of using +IV as an intermediate in some enzymatic transformations), and for Cu, for which +I and +II (possibly +III as an intermediate) are the only relevant oxidation states.

It is notable that for all these redox pairs, the difference in oxidation state is always one electron. Even if nature performs multi-electron redox reactions, such as the oxidation of water to dioxygen, which requires four electrons, the overall reaction is inevitably broken down into one-electron steps. Depending on the reaction path, the electrons are shuttled one-by-one to or from the redox active reaction center. Just in passing, there are of course also redox-active organic molecules, for example flavines, which may take part in redox reactions as cofactors alone or in combination with redox-active metal centers.

While we are all familiar with the electron flow in electrical wires, the question arises how electrons are shuttled to or from the reaction centers of redox-active proteins. In other words, what is the molecular equivalent of an electrical wire in the molecular (or biological, for the purpose of this book) world? The theory of electron transfer was developed by Rudolph A. Marcus, who received the Nobel Price in Chemistry in 1992 for his seminal discoveries. His theory is nowadays well known as the semiclassical theory of electron transfer, or briefly as "Marcus Theory", and it has been amply confirmed by experiment. [1, 2]

In biological systems, electrons "flow" from one redox-active group to the next. In this sense, a "string" of redox active groups with matching potentials placed at the right distance is the molecular equivalent of an electrical wire. Electrons can travel along this molecular wire over distances up to a few hundred Å at significant speed. It is interesting to note that the speed of electron transfer does not have to be close to the speed of light as in electrical wires, nor does it even have to be ultra-fast in every case. However, the systems are optimized by evolution such that the (overall) rate of electron transfer is fast enough for the turn-over frequency of the enzyme – which may only require a few electrons per second in the case of a slow enzyme. In almost all natural systems, electrons are transferred one-at-a-time. Occasionally, nature uses organic cofactors which can undergo reversible one-electron redox reactions in special cases. In the vast majority of cases, however, transition metal containing proteins are used as electron relays: the so-called blue Cu proteins, the heme-Fe containing cytochromes, and the family of Fe–S clusters. All three classes of compounds have been extensively investigated and are described in great detail in common bioinorganic chemistry textbooks, so they will not be covered here in any detail. Just as an example, Figure 8.1 shows

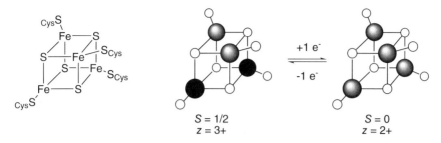

$S = 1/2$
$z = 3+$

$+1\ e^-$
$-1\ e^-$

$S = 0$
$z = 2+$

Fig. 8.1 Left: Schematic drawing of an Fe_4S_4 cluster. The cluster is connected to the protein by cystein sulfur atoms (S_{Cys}) coordinating to the iron atoms. Right: A representation of the one-electron redox chemistry of this cluster. Grey shading of the iron atoms indicates +2.5 charge, black shading indicates +3 charge. Also, the spin state (S) and the overall charge of the cluster is given.

a schematic representation of the four Fe/four S cluster. In this highly symmetrical structure, the iron and sulfur atoms occupy neighboring corners of a cube. The cluster is usually anchored to the protein by cysteine sulfur atoms binding to the Fe atoms as additional ligands. As usual in biology, the Fe atoms may have oxidation states +II and +III. It is thus tempting but false to think that a total of four electrons can be stored in the cluster (four Fe atoms changing their oxidation state by one integer). Even in this cluster, only a one-electron redox reaction occurs, and at all times there are at least two Fe atoms which share an electron and have a formal oxidation state of +2.5 (delocalized valence). The other two Fe atoms can now either be in the +III state (oxidized cluster), or also have delocalized valence +2.5 (reduced cluster), thus effectively performing a one-electron reduction on the cluster with participation of (only) two Fe atoms.[1]

One other interesting feature, which is of importance for our considerations here, arises if we investigate the distance dependence of k_{ET}, assuming a direct electron transfer ("electron tunneling" or "superexchange", as opposed to a hopping mechanism with several discreet intermediates) between donor and acceptor. This distance dependence is exponential as

---

[1] Depending on the redox potential, there are cases where the cluster shuttles between $Fe(2.5)_4$ and $Fe(2.5)_2Fe(2)_2$. Again, this is only a one-electron process and in any given cluster or protein, only this *or* the above redox reaction is realized.

described by the Marcus–Jortner-Levich equation (Eq. (8.1)).

$$k_{ET} = k_0 exp(-\beta R) \tag{8.1}$$

k_{ET} : Electron transfer rate over distance R

k_0 : Electron transfer rate at distance $0\,\text{Å}$

R : Distance (in Å)

β : Exponential factor (dependent on medium, in Å^{-1})

In this equation, the distance dependence is described by the factor β which depends on the medium through which the electron travels. The exponential factor β is $3.5\,\text{Å}^{-1}$ in vacuum, $1.65\,\text{Å}^{-1}$ in water, and around $1\,\text{Å}^{-1}$ in proteins. If we place two FeS clusters at $20\,\text{Å}$ distance, then the rate constants for electron transfer will be 10^{17} years (vacuum), 5×10^4 years in water, but only ms in proteins, respectively. From these facts, two important conclusions arise. First, electron transfer (ET) is only reasonably fast through organic matrices like proteins; water is a much worse medium for ET. Second, even through proteins, ET becomes slow at distances $> 20\,\text{Å}$. As a rule of thumb, it is reasonably fast only for distances $< 15\,\text{Å}$. Indeed, if larger distances must be crossed then nature uses "electron relays" at strategic distances around 10 to $15\,\text{Å}$ and the electron is passed on from metal cluster to metal cluster.

One example is shown in Figure 8.2, which shows the crystal structure of a hydrogenase enzyme. [3] Hydrogenases split molecular dihydrogen (H_2) into protons and electrons. [4] The reaction occurs at the active site (which happens to be a very unusual FeS cluster) deep inside the enzyme. [5] In the particular hydrogenase shown in Figure 8.2, the electrons thus produced leave the active site via a chain of three FeS clusters towards the surface of the protein, where they are taken up by a cytochrome and serve as the source of energy of the organism. The potentials of these clusters are matched according to the Marcus equation such that electron transfer is possible at decent rates. And as indicated in Figure 8.2, the distances between them are $< 15\,\text{Å}$, again making fast electron transfer possible in these enzymes.

Unfortunately, it is beyond the scope of this lab course to isolate and investigate electron transfer proteins or hydrogenases, let alone investigate electron transfer rates. Rather, we will perform some classic electro-chemical experiments on redox-active metal complexes which also have well-defined redox chemistry. For this purpose, we will use organometallic

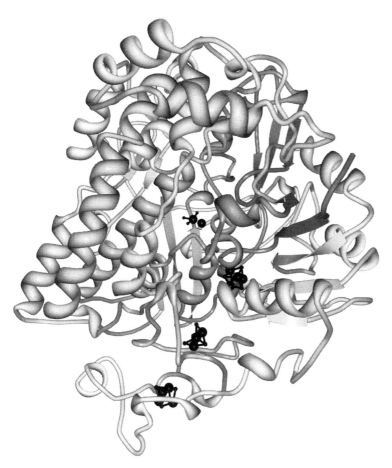

Fig. 8.2 Schematic drawing of the structure of a hydrogenase enzyme showing the chain of three FeS clusters (black) placed at strategic distances of ca. 10 Å from each other which shuttle the electrons from the active site (black, in the middle of the protein) to the surface of the protein. Created from pdb entry 1 FRV [3].

complexes, namely the metallocenes such as ferrocene, ruthenocene, or cobaltocene, which were introduced in Chapters 6 and 7. Ferrocene in particular can, in fact, serve as an artificial "electron relay" in proteins. The chemistry for covalently linking such an artificial electron relay to proteins was described in the previous Chapter 7. Here, we will investigate the electrochemical properties of ferrocene, but also its bioconjugates from the previous two chapters in a little more detail. Beyond the fundamental interest in such studies, as exemplified above for the understanding of electron transfer in enzymes and biological systems in general, there is also significant commercial interest in redox-modified biological systems. For example, a ferrocene-modified enzyme, glucose oxidase, is used in hand-held devices for the quick determination of glucose levels in the blood of patients suffering from diabetes. In these devices, electron transfer would be $> 25\,\text{Å}$ from the active side (again deeply buried inside the enzyme) to the electrode of the device, and thus extremely slow. A suitable ferrocene derivative is attached to the surface of the enzyme, working as an electron relay. By thus breaking the large distance into two roughly equal distances of ca. $15\,\text{Å}$, electron transfer becomes fast and the device a commercial reality. [6]

Metallocenes are a prototypical class of organometallic compounds. [7] For many chemists, the isolation of ferrocene, [8, 9] and its subsequent structure elucidation in 1952 [10, 11] marks the advent of modern organometallic chemistry. [12] Ferrocene itself, or its chemical relatives, are by no means unusual compounds. Yet they offer a combination of properties which makes them attractive not only from an academic point of view, but also for many applications. Among those properties of ferrocene are good chemical stability under a variety of conditions, including exposure to air and water, and ready chemical functionalization which, in combination with its good stability, leads to numerous ferrocene derivatives in the literature. [13, 14] Finally, ferrocene shows a very well-behaved reversible, one-electron oxidation. Indeed, it has been proposed as a secondary standard for electrochemical experiments by IUPAC due to its very favorable electrochemical properties.

The orange-brown ferrocene is readily synthesized from Fe(II) salts like $FeCl_2$ or $Fe(acac)_2$ (acac: acetylacetonate), cyclopentadiene (CpH) and base. It has a symmetrical structure with two parallel cyclopentadienyl rings. The Fe−C distance in ferrocene is $2.06\,\text{Å}$, which corresponds to a distance of $3.32\,\text{Å}$ between the centroids of the two Cp rings. [7, 15] The cyclopentadienide ligand is planar and formally aromatic with six π electrons. [16] All C−C distances are similar at $1.41\,\text{Å}$. The relative orientation of the cyclopentadienyl rings has been the subject of debate: eclipsed with D_{5h} symmetry or staggered D_{5d} symmetry (Fig. 8.3)? In unsubstituted ferrocenes they are in an eclipsed geometry in the gas phase and in the low-temperature ($T < 110\,\text{K}$) orthorhombic crystalline phase. Upon warming, several phase transitions exist that are related to the onset of Cp ring rotation even in the solid state. The barrier of activation has been calculated to be only a few $kJmol^{-1}$, in line with this

staggered
symmetry D_{5d}

eclipsed
symmetry D_{5h}

Fig. 8.3 Eclipsed and staggered conformation of ferrocene.

observation. The weak attractive interaction between the protons on the Cp rings is overridden by steric repulsion in substituted ferrocenes such as the decamethyl ferrocene (Cp_2^*Fe, Cp^* : pentamethyl cyclopentadienyl), resulting in staggered (D_{5d}) conformations in all conditions. [7, 15] In principle, similar considerations apply to most metallocenes, although obviously the structures will differ slightly. The bonding situation in all metallocenes can be readily explained by formally combining the metal d orbitals with two sets of Cp orbitals and filling in the appropriate number of electrons. [17] Most importantly for our discussion here, the border orbitals are only weakly bonding/antibonding and electrons can be added or removed without major distortions of the structure.

In the case of ferrocene, the neutral Fe(II) form has formally 18 electrons and is the most stable species (Fig. 8.4). It can be readily oxidized to form the green ferrocenium cation, which can be isolated as the PF_6^- salt and itself is a mild one-electron oxidant. This one-electron oxidation is reversible under most conditions and the potential is fairly independent of the experimental conditions and highly reproducible. The higher congeners of Fe display a more complicated electrochemistry, with an irreversible two-electron oxidation for ruthenocene and two irreversible one-electron steps for osmocene. Cobaltocene, on the other hand, has 19 electrons in its neutral Co(II) form and is thus a strong one-electron reductant, yielding the very stable cobaltocenium cation, $(Cp_2Co)^+$. In this case, the cation is the most stable form under physiological conditions (Fig. 8.4). Cobaltocene has a redox potential of ca. -1 V compared to ferrocene. Substituents on the Cp ring will have a significant but smaller influence on the redox potential. For example, addition of methyl groups will shift the redox potential of ferrocene about 35 mV to more negative values per methyl group by adding electron density to the ring and effectively stabilizing the oxidized form. Likewise, a carboxyl group

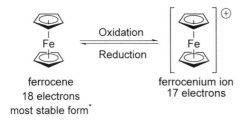

ferrocene
18 electrons
most stable form*

ferrocenium ion
17 electrons

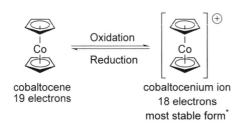

cobaltocene
19 electrons

cobaltocenium ion
18 electrons
most stable form*

Fig. 8.4 Redox chemistry of metallocenes used in this experiment. *"Most stable form" refers to ambient conditions in the chemical laboratory.

removes electron density from ferrocene, making it harder to oxidize and shifting the potential to ca. $+200 \, mV$.

Chemically, the cyclopentadienyl rings in metallocenes behave similarly to aromatic compounds like benzene. [7, 16] Combined with the exceptional stability of the compound, numerous derivatives of ferrocene in particular with different substituents on the Cp rings are easily accessible synthetically. This, combined with the fact that they will mostly have well-behaved, reversible electrochemistry but with slightly different potentials, makes ferrocenes ideal biomarkers for electrochemical detection. Applications of ferrocenes in electrochemical biosensors are as glucose sensors, [6] which are easily used by patients suffering from diabetes to determine the sugar content in their blood, and more recently electrochemical DNA biosensors. [18]

In this experiment, we investigate the redox chemistry of ferrocene in dependence on the concentration and the scan rate. Comparison can be made to the electrochemical properties of other metallocenes, such as ruthenocene and cobaltocene, and derivatives of those compounds. By measuring at different scan rates, different peak currents will be observed. The peak current can be plotted against the scan rate. For electrochemically reversible processes, a linear dependence will be observed. This, in fact, is a very good criterium to establish reversibility.

Furthermore, the electrochemical properties of the metallocene-peptide or metallocene-protein bioconjugates from the previous two chapters can be measured. By evaluating the results from measuring peak currents at different potentials, diffusion coefficients of the conjugates can be determined. You will find that, as expected, larger molecules diffuse more slowly.

Experiment

Objectives

- ✓ synthesis of metalloscenes with different redox properties (optional)
- ✓ use of three-electrode electrochemical set-up
- ✓ qualitative comparison of electrochemical properties
- ✓ determination of peak currents from cycling voltamograms
- ✓ determination of diffusion coefficients from concentration-dependent measurements

Materials

- tetrabutylammonium hexafluorophosphate (TBAPF$_6$, $M = 387.43 \, \text{g} \cdot \text{mol}^{-1}$)
- ferrocene carboxylic acid ($M = 230.05 \, \text{g} \cdot \text{mol}^{-1}$)
- ferrocene ($M = 186.03 \, \text{g} \cdot \text{mol}^{-1}$)
- ferrocenoyl-enkephalin peptide (from Chapter 6)
- ferrocenoyl-lysozyme conjugate (from Chapter 7)
- anhydrous, degassed acetonitrile
- electrochemistry system: potentiostat with electrochemical cell (e. g. Princeton Applied Research Basic Electrochemical System or similar)
- electrodes: Pt wire as counter electrode, "glassy carbon" working electrode (2 mm diameter), Ag/AgNO$_3$ reference electrode

An example of a typical electrochemical cell is shown in Figure 8.5.

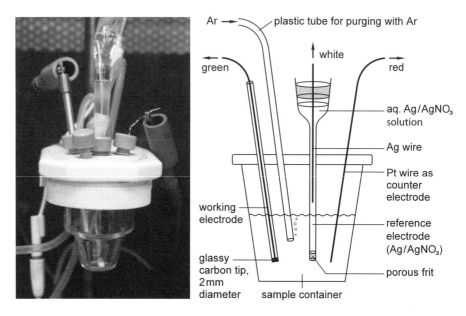

Ar → plastic tube for purging with Ar

↑ white

← green   red →

aq. Ag/AgNO$_3$ solution

Ag wire

Pt wire as counter electrode

working electrode

reference electrode (Ag/AgNO$_3$)

glassy carbon tip, 2 mm diameter

porous frit

sample container

Fig. 8.5 Left: Photograph of the electrochemical cells used with glass vessel and small teflon-coated stirrer bar for solutions to measure, electrodes are connected to color-coded wires. Left – working electrode, middle – reference electrode as a glass capillary, right – Pt wire counter electrode. The tilted plastic tubing behind the electrodes is for purging the vessel with inert gas (Ar, N$_2$). Right: Schematic drawing of the electrochemical cell.

Preparation of the electrochemical cell for measurement

Dissolve 194 mg of TBAPF$_6$ in the electrochemical cell in 4.9 ml of dry, degassed acetonitrile. This 0.1 mM TBAPF$_6$ solution is needed as the supporting electrolyte.

Preparation of a stock solution of the ferrocene derivative

Dissolve a few mg (exact amount calculated from the molecular weight of the sample to be investigated) of your sample compound in a graduated flask in 5 ml of dry degassed acetonitrile. Dilute this solution to obtain a ca. 1 mM solution.

Preparation of the potentiostat

Make yourself familiar with the electrochemical setup to be used. In particular, you will need to be able to adjust the scan range and the scan rate. Mount the electrochemical cell with the solution of $TBAPF_6$ in acetonitrile inside the Faraday cage. Connect the electrodes. We recommend the following color coding: white – $Ag/AgNO_3$ reference electrode; green – glassy carbon working electrode; red – Pt wire counter electrode. Degass for some minutes by bubbling dinitrogen or Ar through the solution. Record a blank cyclic voltammogram (CV) between ca. -1 V to 1 V (scan rate ca. 300 mV/s) of the pure solvent plus $TBAPF_6$. Purging and stirring should always be stopped while recording the CV.

Measuring your sample at different scan rates

Add at least $100 \, \mu l$ of your 1 mM sample (more may be necessary to obtain a good signal-to-noise ratio, depending on your system) solution to ca. 5 ml of the $TBAPF_6$ solution, degass again and record another CV between -1 and $+1$ V at a high (e. g. $1000 \, mV \cdot s^{-1}$) scan rate to obtain an overview. Choose an appropriate potential range (at least $\pm 300 \, mV$ from the peak positions) and record five CVs with different scan rates between 100 and $1000 \, mV \cdot s^{-1}$.

Measuring your sample at different concentrations

Add four times $100 \, \mu l$ each of your 1 mM sample solution to ca. 5 ml of the $TBAPF_6$ solution, degass again and record one CV at the appropriate potential range as determined in the previous experimental step, at each concentration. You can use the scan rate of your choice for this experiment (e. g. $200 \, mV \cdot s^{-1}$), but it must be the same for all concentrations. If you needed to use higher amounts for good signal-to-noise in the first place, then increase concentrations here accordingly.

Calibrating your potential

At the end of your measurements, add one small crystal of ferrocene into the electrochemical cell, let it dissolve and record a CV again at the optimal scan rate after degassing. The new peak which is observed originates from the ferrocene/ferrocenium redox couple. Its potential is set to 0 V by definition and can hence be used to calibrate your results by subtracting its potential from all measurements.

Evaluation of results

Determine $E_{1/2}$ from the last CV (calibration with ferrocene) for the ferrocene/ferrocenium couple. Use this potential (set to $0\,V$ by definition) to calibrate the $E_{1/2}$ potentials from all previous measurements ($E_{1/2,\text{corr.}} = E - E_{1/2}(\text{FcH}/\text{FcH}^+)$).

Determine the cathodic and anodic peak currents for all measurements graphically as indicated in Figure 8.6. When you plot your data, make sure you follow the conventions for representation of electrochemical data as indicated in Figure 8.6: Oxidative potential to the right, and anodic current up.

Plot cathodic and anodic peak currents from the scan rate-dependent experiments at a given concentration against the square root of the scan rate and carry out a linear regression analysis.

Plot cathodic and anodic peak currents from the sample concentration-dependent measurements at a given scan rate against the sample concentration and carry out a linear regression analysis.

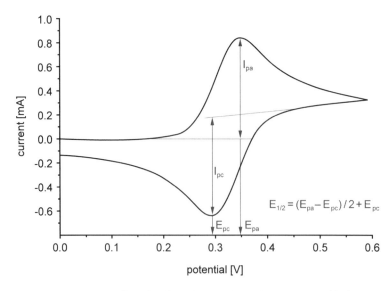

Fig. 8.6 Example for plotting the cyclic voltamogram (CV) and determining peak currents.

Data are analyzed according to the Randles–Sevcik equation (Eq. (8.2)). This equation also provides a criterium for reversibility, i. e. if the peak current follows the square root of the scan rate, the reaction is electrochemically reversible. From the slope of the above two curves, you should be able to calculate the diffusion coefficient of your compound.

$$i_p = 2.69 \times 10^5 n^{3/2} v^{1/2} D_{red}^{1/2} c_{red} \tag{8.2}$$

i_p : peak current
n : Number of electrons transferred (one for ferrocene)
v : scan rate
D_{red} : Diffusion coefficient
c_{red} : Concentration of the compound

Variations of the experiment

All of the above experiments can be carried out on the other metallocenes (ruthenocene and cobaltocenium hexafluorophosphate). Note that the Randles–Sevcik equation does not necessarily hold in these cases.

You can also investigate different metal bioconjugates, such as the ferrocenoyl-peptide and ferrocenoyl-protein from the previous two chapters. Both compounds should diffuse much slower than ferrocene alone. In principle, such experiments may be carried out with any metal complexes or metal-conjugates. In our experience, however, ferrocene conjugates usually work best because the electrochemistry is clean and reversible.

Additional questions

✓ Draw the molecular orbital scheme for ferrocene. Discuss the bonding situation! From which orbital will the electron be removed during electrochemical oxidation?

✓ What is the purpose of the TBAPF$_6$ in this experiment (also called the supporting electrolyte)?

✓ Why do you have to use a reference electrode? What is its purpose? Why is ferrocene added at the end of a series of measurements and its potential determined?

Bibliography

[1] J. R. Winkler, H. B. Gray, Electron-transfer in ruthenium-modified proteins, *Chem. Rev.* **1992**, 92, 369–379.

[2] H. B. Gray, J. R. Winkler, Electron transfer in proteins, *Annu. Rev. Biochem.* **1996**, 65, 537–561.

[3] A. Volbeda, M. H. Charon, C. Piran, E. C. Hatchikian, M. Frey, J. C. Fontecilla-Camps, Crystal-structure of the nickel-iron hydrogenase from *Desulfovibrio gigas*, *Nature*, **1995**, 373, 580–587.

[4] R. H. Morris in *Concepts and Models in Bioinorganic Chemistry*, H. B. Kraatz, N. Metzler-Nolte (eds.), Wiley-VCH, Weinheim, **2006**, 331–362.

[5] D. J. Evans, C. J. Pickett, Chemistry and the hydrogenases, *Chem. Soc. Rev.*, **2003**, 32, 268–275.

[6] J. Wang, Glucose biosensors: 40 years of advances and challenges, *Electroanalysis* **2001**, 13, 983–988.

[7] N. J. Long, *Metallocenes*, Blackwell Science, Oxford, **1998**.

[8] T. J. Kealy, P. L. Pauson, *Nature* **1951**, 168, 1039–1040.

[9] S. A. Miller, J. A. Tebboth, J. F. Tremaine, Dicyclopentadienyliron, *J. Chem. Soc.* **1952**, 632–635.

[10] E. O. Fischer, W. Pfab, Cyclopentadien-Metallkomplexe, ein neuer Typ metallorganischer Verbindungen, *Z. Naturforsch.* **1952**, 7*b*, 377–379.

[11] G. Wilkinson, M. Rosenblum, M. C. Whiting, R. B. Woodward, The structure of iron bis-cyclopentadienyl, *J. Am. Chem. Soc.* **1952**, 74, 2125–2126.

[12] G. Wilkinson, The iron sandwich. A recollection of the first four months, *J. Organomet. Chem.* **1975**, 100, 273–278.

[13] A. Togni, T. Hayashi, *Ferrocenes: Homogeneous catalysis, organic synthesis, material science*, VCh, Weinheim, **1995**.

[14] P. Stepnicka, *Ferrocenes*, John Wiley & Sons, Inc., New York, **2008**.

[15] C. Elschenbroich, *Organometallics*, Wiley-VCh, Weinheim, **2006**.

[16] R. B. Woodward, M. Rosenblum, M. C. Whiting, A new aromatic system, *J. Am. Chem. Soc.* **1952**, 74, 3458–3459.

[17] A. Haaland, Molecular structure and bonding in the 3d metallocenes, *Acc. Chem. Res.* **1979**, 12, 415–422.

[18] D. R. van Staveren, N. Metzler-Nolte, Bioorganometallic chemistry of ferrocene, *Chem. Rev.* **2004**, 104, 5931–5985.

9 Metal complexes with anti-proliferative activity

Summary. In this chapter, you will be introduced to metal complexes which are used in the treatment of diseases. This will be exemplified by the drug Cisplatin, which is one of the most widely used anti-cancer drugs in the clinic. You will learn about the mechanism of action of metal-based anti-cancer drugs and drug candidates. Since clinical studies in humans are for ethical reasons only possible at a very late stage of drug development, cell-based assays are used to evaluate the anti-proliferative activity of any new compound. In the experimental part, you will prepare metal complexes with anti-proliferative activity and test their activity in such a cell-based colorimetric assay in a qualitative and quantitative way. The properties of several such assays will be discussed.

Learning targets

✓ Metal complexes as anti-cancer drugs and drug candidates
✓ Biological targets of metal-based anti-cancer drugs and drug candidates
✓ Synthesis of metal complexes with anti-proliferative activity
✓ Cell-based biological assays to evaluate the anti-proliferative activity of metal complexes
✓ Determination of IC_{50} values from cell-based assays

Background

In 1969, the group of Barnett Rosenberg reported that tumor-bearing mice were cured from the tumor by a very simple inorganic compound, namely *cis*-diammino-dichloro-platin(II) (Fig. 9.1). [1] The discovery became a milestone in medicinal chemistry and the compound, called Cisplatin for short, was quickly submitted to clinical trials and was approved for the treatment of cancer by the American Food and Drug

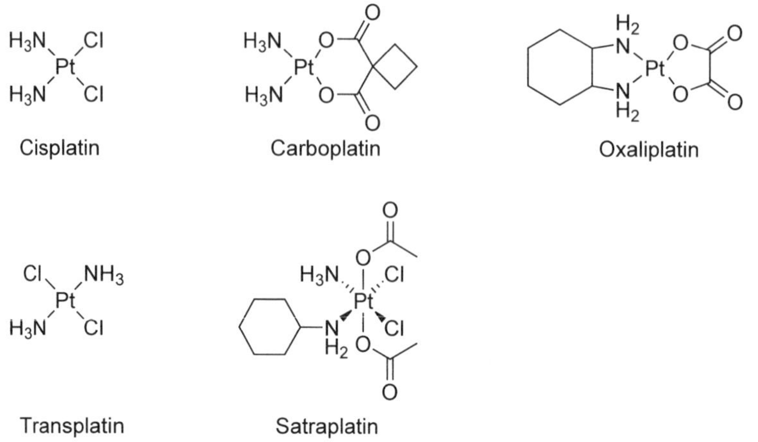

Fig. 9.1 Platin-based anti-cancer drugs and drug candidates.

Administration (FDA) in 1978. Cisplatin became an immensely useful drug, and it is still among the three most frequently used drugs for the treatment of a variety of cancers. As the most spectacular example, testicular cancer (for which no useful therapy existed previously) became curable with > 90% probability when treated with Cisplatin. For many other types of cancer, Cisplatin is used together with other drugs and it has a profound ability to extend the life span of patients, and also improve their quality of life. Following the success of Cisplatin, literally thousands of other platin compounds were synthesized and tested for their anti-proliferative potential. Out of those, two platinum-based drugs are approved worldwide (Carboplatin and Oxaliplatin, see Figure 9.1), and a few more gained regional approval. Cisplatin, like all other approved Pt anti-cancer drugs so far, is a Pt(II) compound. In recent years, octahedral Pt(IV) compounds received considerable attention because they have different pharmacokinetic properties and a different profile of activity. Inside the tumor cells, they are probably reduced to the square planar Pt(II) complexes with loss of the two axial ligands. One Pt(IV) compound, Satraplatin (see Fig. 9.1), was very close to approval as the first orally available Pt anti-cancer drug and it has only recently failed late clinical phase III trials. Interestingly, the *trans* enantiomer of Cisplatin (also called Transplatin, Fig. 9.1) is at least one order of magnitude less active than Cisplatin and therefore has never entered clinical trials. [2, 3]

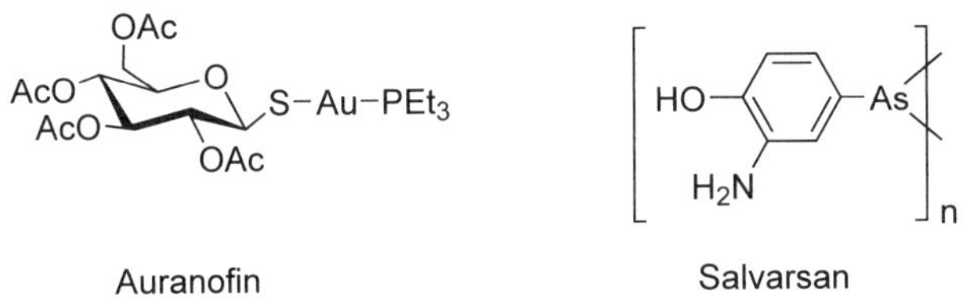

Fig. 9.2 Auranofin and Salvarsan, two historically important metal-based drugs.

On the other hand, the success of Cisplatin has not just lead to the development of new Pt-based drug candidates. It has also provided a major boost in all different research efforts to explore the potential of metal-based drugs. This field, which has become very diverse and forms a research area in its own right, is known as *medicinal inorganic chemistry*. [4,5]

It should be remembered, however, that Cisplatin is not the first inorganic compound to have significance in medicinal chemistry, nor is it the only one with clinical approval. The use of metals for the treatment of diseases dates back probably thousands of years, starting with the use of gold in several ancient cultures, albeit in more mystical procedures. While wonderous effects are still associated with this metal, e. g. in the form of colloidal drinks of gold ("aurum potabile"), Au-based drugs are, in fact, used nowadays for the treatment of rheumatoid arthritis (Auranofin, Fig. 9.2). Already several centuries ago it was recognized that iron salts are helpful for the treatment of anemia. Today, we understand that iron deficiency in the human body is relatively common and can be readily and cheaply treated by supplementing iron in the form of its +II salts. Lithium carbonate is an important compound for the treatment of bipolar disorders such as maniac depression. A few other compounds also contain metal centers and are in use for special cases. Interestingly, the very first systematic screening of compounds for medicinal activity was carried out by a German chemist, Paul Ehrlich, at the beginning of the 20th century. Ehrlich had synthesized several hundred arsenic compounds and subjected them directly to animals which were infected with syphilis, a severe and at that time incurable disease. One of the compounds, named Ehrlich no. 606, turned out to be highly effective against the disease, and it was soon marketed under the name of Salvarsan (Fig. 9.2). Interestingly, the proposed chemical structure of Salvarsan is somewhat simplified. Although this was a successful drug for several decades, there is no

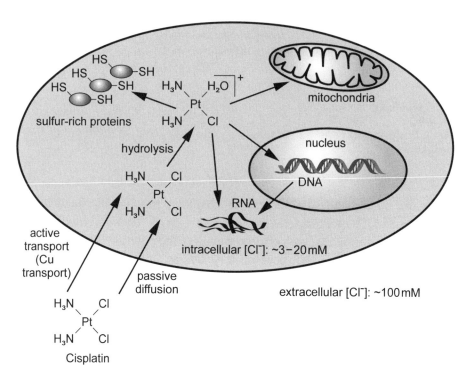

Fig. 9.3 Proposed mechanism of action of Cisplatin.

crystal structure of Salvarsan available. A recent mass spectrometric study indicates a dynamic mixture of several small ring isomers in solution. [6]

With the advent of modern biochemistry and cell biology, drug research and drug discovery have changed from a trial-and-error-type screening to more systematic efforts. At the heart is the identification of the molecular, cellular target of a drug. In modern drug discovery, identification of a viable, "drugable" target, e. g. an enzyme, would precede chemical efforts to synthesize drugs that interact with this target, in our examples inhibitors of the identified enzyme. [7] To understand the mechanisms of older drugs, it is important to identify their biological targets, which in most cases will require identification of the molecules in the cell that interact with that specific drug. These could be enzymes, but also other cellular structures like outer or inner cell membranes, RNA, or DNA in the cellular nucleus. For the anti-cancer drug Cisplatin, DNA has been established as the ultimate target (Fig. 9.3). [8, 9]

In the traditional model, Cisplatin gains entry into cells in its intact, uncharged form because the extracellular chloride ion concentration is relatively high (ca. 100 mM). Once inside the cell, where the chloride ion concentration is only a few mM, one chloride ion dissociates rapidly from the complex, and the second more slowly. The resulting charged Pt complex is thereby trapped inside the cell. More recent work questions this model, which is based purely on diffusional uptake. The involvement of Cu ion transporter systems is proposed, which serve to actively incorporate the drug into the cells. Likewise, an active excretion system for Cu ions exists, which is believed to be involved in resistance to Cisplatin. In any case, the charged Pt species interact with proteins, in particular the soft cysteine sulfur atoms. [9, 10] For example, glutathione, which is an important redox mediator in the cells, is present in mM concentrations and has been shown to interact with Cisplatin-derived cations. These interactions, however, are reversible. Ultimately some of the Pt ions will enter into the cellular nucleus and reach the DNA. Here again, many modes of interaction are conceivable. It has been shown that the most nucleophilic sites on DNA are the guanine N7 atoms, and to a lesser extent the adenine N7 atom. If the Cisplatin dication interacts with two neighboring guanines, very stable adducts are formed. These guanine bases can be on the same or opposite strands of DNA, and the adducts are hence denoted intra- or inter-strand. In either case, these lesions cannot be repaired by the cellular DNA repair machinery. If too many such lesions are present in one single cell, then a signal is sent as a consequence of futile repair efforts, causing the cell to commit suicide in a regulated, controlled fashion. This process of "cellular suicide" is called apoptosis, and commonly it is assumed that a minimum of at least 1000 Cisplatin-DNA intra- or inter-strand adducts are required to initiate apoptosis. [11] Apoptosis itself is a complicated process, in which caspase enzymes play a key role, cytochrome c is released from the mitochondria, the nuclear DNA is fragmented, and ultimately the cells shrink and are disposed of by the body. While the fragmentation of DNA can be easily detected at the very late stages of apoptosis (a so-called "DNA ladder" is seen in an electrophresis gel), phosphatidylserine is an important early marker for apoptosis. This particular molecule is irreversibly turned to the outside of the cell membrane, where it can be detected colorimetrically by the Annexin V assay. The induction of apoptosis is an important and desirable feature of anti-proliferative drugs as it really kills tumor cells ("cytotoxic"), rather than just inhibiting tumor cell growth, which can be resumed once the concentration of the drug decreases again ("cytostatic").

This mechanism of action, and particularly the importance of DNA interaction, has developed almost into a paradigm for *any* metal-based anti-cancer drugs. For many years, DNA interaction was inevitably and without asking assumed as *the* mechanism of action for *any* cytotoxic metal complex. This view is changing slowly only in recent years, with researchers considering different mechanisms of action for chemically and structurally different metal complexes.

An unbiased look at the periodic table reveals an enormous amount of possibilities for medicinally active metal complexes. It is not just the fact that there are more than 50 different metals (compared to the six elements that make up most organic molecules: C, H, O, N, P, S). Also, a greater structural variety is possible in principle: Whereas a tetrahedral carbon center can give rise to two stereoisomers, an octahedral metal center can have up to 30 different stereoisomers! Moreover, metal centers may possess particular reactivity (such as redox activity). When combined with organic ligands, these particular properties can be transferred into target-selective molecules for which well-established rules in medicinal (organic) chemistry exist. [12]

Organometallic compounds offer particular promise in this respect because they are kinetically inert, often uncharged and fairly lipophilic. The metal centers exist in low oxidation states, which limits the danger of oxidative damage inside cells. Finally, a broad range of structural and functional variations is available through (organic) synthetic chemistry. Figure 9.4 lists a few organometallic compounds, which are seen as promising lead structures for anti-cancer drugs. [13–16]

It is interesting to note the structural variety of the complexes in Figure 9.4. Even the very simple compound ferrocene was found to have some anti-proliferative activity, albeit only at relatively high concentrations. [17] Its mechanism of action remains elusive despite several efforts, however, it is believed by most researchers that redox chemistry (Fe(II) $\longleftrightarrow$ Fe(III), see Chapter 8) is at the heart of its activity. [18] Another metallocene, namely titanocene dichloride, has been in clinical trials until recently. Its failure to prove effective can probably be attributed to its instability in aqueous medium and concurrent problems with an adequate formulation of the compound as a drug. A derivative of titanocene dichloride, termed titanocene Y, represents the second generation of metallocene anti-cancer drugs and has far better activity than titanocene dichloride itself. For

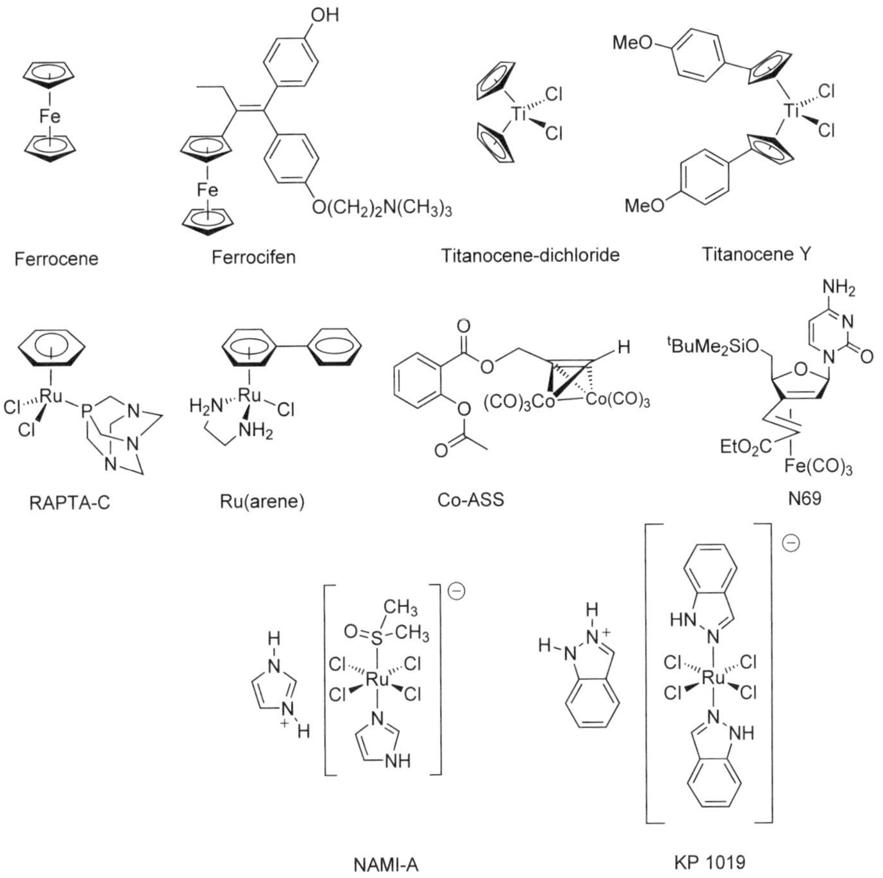

Fig. 9.4 Some organometallic compounds with promising anti-cancer properties: ferrocene, ferrocifen, titanocene-dichloride, titanocene Y, RAPTA-C, arene-Ru compound, Co-ASS, and N69 (an iron carbonyl nucleoside), NAMI-A und KP 1019.

both compounds, however, the exact mechanism of action is unclear at present. [19] Quite on the contrary, the ferrocene derivative ferrocifen was designed after a well-known drug against breast tumors, tamoxifen. Like tamoxifen, ferrocifen is active against breast tumor cell lines which express the estrogen receptor (ER(+) cell lines). Unlike tamoxifen, however, ferrocifen is also active against ER(−) breast cancer cell lines, which do *not* express the receptor. This is a very significant finding because this latter class represents about one third of all tumors and is far more difficult to treat by chemotherapy than ER(+) tumors. [20, 21] Again, redox chemistry of ferrocene has been convincingly demonstrated to be responsible for activity, [22] and a new generation of structurally unrelated ferrocene-derived drug candidates has been proposed. [23] Two metal carbonyl complexes (Co-ASS and N69) with very promising anti-cancer activity are also included in the above Figure 9.4 to demonstrate the structural variations. [16] Finally, the Ru compounds RAPTA-C and NAMI-A in Figure 9.4 are interesting because they are not working anti-proliferative against the primary tumor, but they are anti-metastatic. [24–26] Indeed, the majority of cancer patients do not die from the primary tumor, but from complications in the disease which are caused by metastases. Very often, these metastases are more difficult to treat by chemotherapy because they have already acquired some resistance against anti-cancer drugs. On the other hand, the process of metastasis is poorly understood, and there is really no general concept yet for combating metastasis. The two complexes NAMI-A and RAPTA-C fall into this category of anti-metastatic drugs. Although not organometallic, NAMI-A and also KP1019 were included in this figure because both are in clinical trials and are thus amongst the clinically most advanced inorganic drug candidates.

Preclinical screening of potential tumor therapeutics is an important field of research, because it can already show the potential for a cytotoxic compound and thus help reduce the cost and time for further development. One of the screening methods routinely used are human tumor in vitro cell line assays, in which potential drug candidates are screened in a comparatively short time against up to 60 representative human tumor cell lines. [27] This procedure is relatively easy and can be adopted on a small scale for lab courses to visualize the anti-proliferative activity of (newly) synthesized metal complexes. In this practical course, we use Cisplatin and some of its derivatives as metal complexes. As the cell line of choice we use the HeLa cell line, which is widely available, fast growing and easy to cultivate. In fact, HeLa cells were the first immortal cell line to be developed for permanent cell culture in medicinal research. They were derived from the cervical cancer of Henrietta Lacks, who died of the disease in 1951. This cell line (and subsequently many others) has undoubtedly had an enormous impact in medicinal and cell biology research, and very likely saved many lives or helped to prolong the life time of others. Yet the fact that these cells were taken from the patient and later commercialized without her explicit consent raises important ethical questions that were also considered in court. [28]

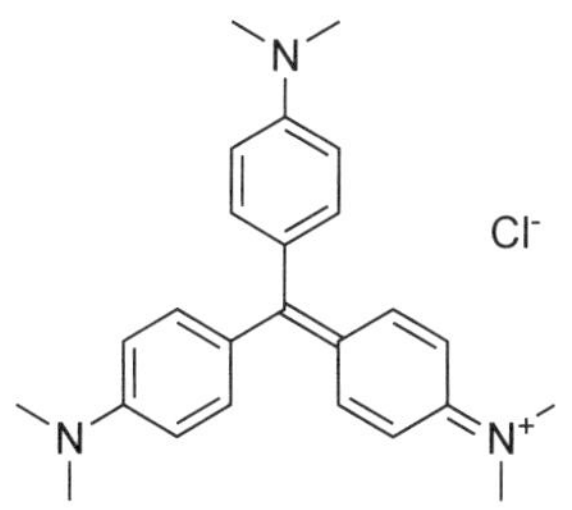

Fig. 9.5 The chemical structure of crystal violet (CV), which is used for colorimetric quantification of the cell biomass in this experiment.

In a laboratory setting it is necessary to determine the anti-proliferative activity of a compound by a quick and reliable assay. In most cases, colorimetric assays are used for this purpose. A color change relates to either the cell number, their biomass, or the metabolic activity of the cells. Treated cells will have fewer cells (or less biomass) in a given volume after some time of treatment, or the present cells are dying/dead and thus have low/no metabolic activity. In this lab experiment, we use crystal violet (CV, see Fig. 9.5) as a dye which interacts with chromatin in cells. The cells are fixed to their well after exposure to the anti-cancer drug candidates, and then CV is added. After washing, only the CV bound to the cells remains and this can be easily quantified colorimetrically. This assay thus measures biomass, which is believed to correlate with cell numbers. The found biomass is compared to an untreated control. Evidently, the underlying principle states that the more cytotoxic a drug is, the smaller the number of cells after the time of treatment. However, it is sometimes more adequate to consider the viability of cells, rather than their number, as in a given experiment many cells may be present, but they are all dead. To this end, formazan dyes were developed, which are reduced by viable (but not by dead) cells, thus undergoing a colour change. The classical MTT or XTT assay is such a metabolic assay. More recent (and experimentally easier) versions use dyes like sulforhodamine B (SRB assay) or resazurin. In all cases, the experiment is performed in multi-well plates (6, 24, 96, 384 or even more wells, depending on the cell numbers and volumes needed) and the results are recorded most easily on a multi-well plate reader that scans the absorption at a given wavelengths automatically in each well.

Experiment

Objectives

✓ preparation of cytotoxic Pt complexes (optional)
✓ use of colorimetric, cell-based crystal violet assay to determine cell biomass
✓ determination of IC_{50} values from a cell-based assay for the HeLa cell line

Materials

– crystal violet (hexamethylpararosaniline chloride, methyl violet 10B)
– paraformaldehyde (PFA)
– ethanol
– disodium hydrogenphosphate, Na_2HPO_4 ($M = 141.96\,g\cdot mol^{-1}$)
– sodium dihydrogenphosphate, NaH_2PO_4 ($M = 119.78\,g\cdot mol^{-1}$)
– sodium chloride, NaCl ($M = 58.44\,g\cdot mol^{-1}$)
– potassium chloride, KCl ($M = 74.55\,g\cdot mol^{-1}$)
– stock solution of the test compound(s), ideally ca. 10 mM in aqueous buffer at pH 7.4
– ultrapure water
– graded volumetric flasks (10, 100, 1000 ml)
– sterile pipettes and tips (10, 100, 1000 μl)
– flat bottom microtiter plates
– gelatin (bovine, if not pre-treated MTPs are used)
– UV/Vis microplate reader
– microplate shaker
– cell culture equipment (sterile bench, incubator, suction device)
– Triton X-100
– if needed as a positive control (or test substance): Cisplatin, $cis-(NH_3)_2Cl_2Pt$ ($M = 300.05\,g\cdot mol^{-1}$)
– light microscope with a Neubauer or Thoma chamber to determine cell numbers
– HeLa cells
– DMEM medium

- fetal calf serum (FCS)
- trypsin
- penicillin
- streptomycin
- L-glutamine
- EDTA

Cell culture conditions

To perform in vitro cytotoxicity assays a cell culture lab is needed which should contain a sterile laminar flow working bench and a CO_2 cell culture incubator. All tools and solutions should be sterilized either by heat or vapor or filtrated through a sterile filter. Prior to perform the cytotoxity assay the cell lines should be kept for a week in culture. Most adherent cell lines can be used, however the assay should be adjusted for the generation time of the cell line. Different viability or chemosensitivity assays can be used to assess the cytotoxicity of metal complexes. Here we will introduce the crystal violet (CV) assay, which is cost-effective and can be continued on different days without the requirement for sterile conditions at all times. As an example we will use the HeLa cell line, which is widely available, fast growing and easy to cultivate. For HeLa cells, DMEM-Medium supplemented with 10% FCS (Fetal Calf Serum) and 2 mM L-glutamine, 100 U/ml penicillin and 100 μg/ml streptomycin should be used as cell culture medium. They should be grown in a humidified atmosphere with 10% CO_2 until 70 to 100% confluence is reached. Upon reaching confluence the cells should be passaged by trypsination with a trypsin 0.25%/EDTA 0.02% solution and plated out in a new cell culture vessel.

Crystal violet cytotoxicity assay

The CV assay determines the cell biomass, which is taken as an (indirect) measure of the cell number. One of the advantages of the CV assay is that it is not a single end-determination measurement like many of the viability assays are. Determination of a start point enables additional control of experimental procedure and the differentiation between cytocidal or cytostatic effects. [29] Crystal violet stains mainly chromatin and is thus a dye for the determination of cell biomass. Read-out is obtained by measuring the UV absorbance at 570 nm.

Preparations for CV assay

Coat the wells of 96-well microtiter plates (MTPs) with $100 \, \mu l$ sterile 0.2% gelatin solution, which is removed directly. MTPs are left to dry under the sterile bench for at least 1 h. Alternatively, a poly-lysine solution or pre-treated MTPs (e. g. Cell+, Sarstedt) can be used.

Cells are trypsinized and resuspended in cell culture medium. Then the cell number of the suspension is determined by a Neubauer or Thoma Chamber. The cell number should be diluted to $40\,000$ cells/ml with cell culture medium. $100 \, \mu l$ of the diluted solution should be filled into each well (seeding), giving 4000 HeLa cells in each well. In addition to the MTPs for the substance testing, a control plate for the determination of the start point value should be seeded. After the seeding of the cells, the MTPs should incubate for 24 h before applying the test substances. Stop the growth of the cells on the control plate after 24 h and apply the procedure described under "execution of the CV assay".

Preparation of the crystal violet staining solution

$40 \, mg$ of crystal violet are first dissolved in 4 ml of ethanol and then filled up to 100 ml with water.

Preparation of $1\times$ PBS (phosphate buffered saline)

Dissolve $8.0 \, g$ sodium chloride, $0.2 \, g$ potassium chloride, $1.44 \, g$ disodium hydrogenphosphate and $0.24 \, g$ sodium dihydrogenphosphate in 800 ml of ultrapure water and adjust the pH as needed (usually to 7.4) with HCl/NaOH. Add water to 1 liter.

Preparation of the fixation solution

The fixation solution is needed to fix the cells to the plate before adding the staining solution. Weigh 4 g of paraformaldehyde and add 100 ml of $1\times$ PBS, add few drops of $0.1 \, \text{M}$ NaOH and heat up to $60\,°C$. Stir until it dissolves. Store at $4\,°C$ until use.

Substance concentration for IC_{50} value determination

Generally substances should be prepared in stock solutions. If their solubility in aqueous media is low, organic solvents like DMSO, methanol and ethanol can be used to a final concentration of 1 to 2% in the incubation medium without doing much harm to the cells. Other solvents can also be used, but will cause greater damage. If the compounds are very acidic the solutions should be buffered to a slightly basic pH. If the substance is soluble in water it can be diluted directly to the desired concentration in cell culture medium.

The IC_{50} values (inhibitory concentration at 50% of cell growth) are calculated from a dose response curve, which should follow a sigmoidal function. To get the shape of a sigmoidal curve from the results, it is necessary to choose the concentration of the test compound around the turning point of the curve. For Cisplatin, a range from 0.1 to 50 μM would be suitable given an IC_{50} value of approximately 10 μM after 48 h. If it is unknown whether the compound exerts cytotoxic properties, and at which concentrations, higher concentrations and possibly a larger range should be used. Once a rough estimate for the IC_{50} has been obtained this way, a more precise determination as detailed below can be carried out in a second experiment. It is recommended to use at least six different concentrations within two to three orders of magnitude to achieve a reasonable sigmoidal fit.

Execution of the CV cytotoxicity assay

After 24 h of cell growth, the substance to be tested can be applied. For this purpose the cell culture medium has to be pipetted away and 100 μl of the different compound concentrations in fresh cell culture medium can be added to the wells of the MTP. If the compound precipitates in aqueous media, alternatively 99 μl of fresh cell culture medium and 1 μl of the stock solution can be added. It should be noted, that in addition to the concentrations of the compounds, non-treated wells and wells which are treated only with the solvent which is used for delivery should be present on the MTP. Incubate the cells for 48 h in order for the effect of the compounds to develop. Following the incubation the cell culture medium is removed and the cells are rinsed with PBS, which is also removed. Then the fixation solution can be added and kept for 20 min at room temperature in the wells. Once the cells are fixated and stick firmly to the well bottom they can be handled easily and working under sterile conditions is no longer necessary. Cells are then washed with 0.1 % Triton X-100 solution in PBS and then rinsed again with PBS. Then 100 μl of the staining solution is added to the wells and left to shake gently for 30 min. After removal, the wells must be rinsed three times with water and then left standing for 10 min filled with water and afterwards dried. Finally, the crystal violet can be extracted by filling 100 μl of ethanol into the wells and gently shaking the MTPs for 3h. The absorption of individual wells are read out in an microplate reader at a wavelength of 570 nm.

Evaluation of results

First, the mean values of all matching wells are calculated. The calculation of the remaining percentual cell mass (CM%) for each concentration follows formula (9.1):

$$CM\% = \frac{(S - I)}{(C_{24h} - I)} * 100 \tag{9.1}$$

S : mean absorbance in wells at specified substance concentration
I : mean absorbance in wells on separate plate at the beginning of substance incubation
C_{24h} : mean absorbance in wells with untreated cells or vehicle control if appropriate
If $S < I$, negative values are possible, stating a cytocidal effect.

These CM% values are then plotted on the ordinate against logarithmical substance incubation on the abscissa. The data can then undergo a fit procedure using a sigmoidal fit function fit or a Boltzmann function fit. Origin (Originlab, USA) is very convenient for this purpose, but also Prism (Graphpad, USA) can be used, as well as many other scientific graph plotting programs. If the IC_{50} value (which is the turning point of the curve) is not directly printed by the software, it can be read from the fitted curve by drawing two orthogonal lines from CM% = 50 to the curve and from there to the abscissa. A sample curve is provided in Figure 9.6.

Variations of the experiment

In this experiment, any metal complex can be tested, provided that sufficient solubility in water/aqueous buffer can be achieved. If a synthetic component to the lab experiment should be added, then the classic compounds Cisplatin ([30], an additional recrystallization from 0.1 M aq. HCl is recommended) and Transplatin [31] (cis-diammino-dichloro-platin(II) and trans-diammino-dichloro-platin(II)) can be easily synthesized and used in the subsequent cell testing. However, it should be kept in mind that these compounds are very toxic and potentially mutagenic. Therefore, great care must be exerted at all stages of manipulation, and

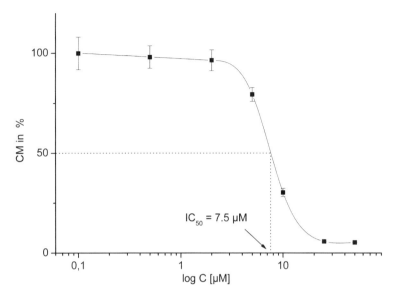

Fig. 9.6 Sample curve from the CV cytotoxicity assay, indicating manual determination of the IC_{50} value on the abscissa.

the synthesis should only be carried out by experienced students under proper supervision.

Many variations are possible with this experiment. In addition (or alternatively) to the CV assay, other assays could be performed. Formazan dyes are widely used in this context, which are reduced by cell metabolites, which in turn induce a color change that can be detected by an increase or decrease in absorbance. Some of these formazan dyes (WST-1 and Resazurin) can even be combined with the CV assay, thus not requiring a second set of microtiter plates. They are simply applied before the CV assay, because they rely on living cells. A review of the different formazan dyes can be found on Roche's Apoptosis and Proliferation manual which is freely available online (http://www.roche-applied-science.com/PROD_INF/MANUALS/CELL_MAN/cell_toc.html). Another commonly used assay is the Sulforhodamine B (SRB) assay. This assay needs acidic fixation of the cells and relies on the unspecific binding of SRB to cell proteins. [32] Furthermore, the incubation time can be varied to assess whether the IC_{50} value of the drug increases or decreases. Values of 24–48 h are appropriate for the experiment described here, but 12–96 h can be used in special cases and cell lines. For Cisplatin, which inter alia inhibits cell division, a prolongation of the incubation time leads to decreasing IC_{50}, while metabolically active drugs like 5-Fluorouracil could lead to higher IC_{50}. Also time-dependent plotting of IC_{50} values can be an option. An additional variation would be use of different cell lines, if available. Drug resistant cell lines exist, but also breast cancer or pancreas cancer cell lines, which normally show a higher resistance to cytostatics, could be used.

Additional questions

✓ Biologically active compounds (be it anti-tumor agents or anti-microbially active compounds, for example) are often classified according to their mode of action. In this sense, Cisplatin is often named together with other "alkylating agents" by medicinal chemists to classify its mode of action. Discuss the pros and cons of this categorization: Which other (organic) compounds are named as alkylating agents? What is their mode of action, what is their primary target, in which way are they similar or different from Cisplatin in this respect?

✓ As described above, Cisplatin has been on the market for a considerable time, and it is evidently an eminently successful drug. However, FDA approval procedures have changed since the introduction of Cisplatin to the clinics. Discuss which steps are required today for FDA approval of a new drug. How would Cisplatin perform in these required steps, if it were submitted for approval today? Which essential facts about Cisplatin are well known? Is there FDA-required information, that is even today not (yet) available for Cisplatin?

✓ As a sideline to the above, it has been stated that Cisplatin would never gain FDA approval as a new drug nowadays (a fate it shares, in the opinion of some scientists, with another classic: Aspirin). Identification of the active metabolite and the exact pharmacological fate of the drug are considered crucial. Compare Cisplatin to other metal-based drugs in this respect: Is this particular to Cisplatin, or do similar considerations apply to other metal-based drugs as well? Why is it generally more difficult to identify active metabolites of a metal coordination compound than of an organic drug in the body? Which techniques can (not) be used? Identification of such issues, and addressing them with scientific rigor, will be essential to change the thinking of medicinal chemists, including those involved in regulatory issues, towards metal-based drugs.

Bibliography

[1] B. Rosenberg, L. VanCamp, J.E. Trosko, V.H. Mansour, Platinum compounds: A new class of potent antitumor agents, *Nature* **1969**, 222, 385–386.

[2] M.A. Jakupec, M. Galanski, V.B. Arion, C.G. Hartinger, B.K. Keppler, Antitumour metal compounds: More than theme and variations, *Dalton Trans.* **2008**, 183–194.

[3] M. Galanski, M.A. Jakupec, B.K. Keppler, Update of the preclinical situation of anticancer platinum complexes: Novel design strategies and innovative analytical approaches, *Curr. Med. Chem.* **2005**, 12, 2075-.

[4] P. Sadler, Medicinal inorganic chemistry, *Angew. Chem. Int. Ed.* **1999**, 38, 1512–1531.

[5] M. Gielen, E.R.T. Tiekink, *Metallotherapeutic Drugs & Metal-based Diagnostic Agents*, John Wiley & Sons, Ltd., Chichester, **2005**.

[6] N.C. Lloyd, H.W. Morgan, B.K. Nicholson, R.S. Ronimus, The composition of Ehrlich's salvarsan: Resolution of a century-old debate, *Angew. Chem. Int. Ed.* **2005**, 44, 941–944.

[7] S.P. Fricker, Metal based drugs: From serendipity to design, *Dalton Trans.* **2007**, 4903–4917.

[8] E.R. Jamieson, S.J. Lippard, Structure, recognition, and processing of cisplatin-DNA adducts, *Chem. Rev.* **1999**, 99, 2467–2498.

[9] D. Wang, S.J. Lippard, Cellular processing of platinum anticancer drugs, *Nature Rev. Drug Discov.* **2005**, 4, 307–320.

[10] J. Reedijk, Why does cisplatin reach guanine-N7 with competing S-donor ligands available in the cell?, *Chem. Rev.* **1999**, 99, 2499–2510.

[11] T. Boulikas, M. Vougiouka, Cisplatin and platinum drugs at the molecular level, *Oncol. Rep.* **2003**, 10, 1663–1682.

[12] W.H. Ang, L.J. Parker, A.D. Luca, L. Juillerat-Jeanneret, C.J. Morton, M.L. Bello, M.W. Parker, P. Paul J. Dyson, Rational design of an organometallic glutathione transferase inhibitor, *Angew. Chem. Int. Ed.* **2009**, 48, 3854–3857.

[13] C.S. Allardyce, A. Dorcier, C. Scolaro, P.J. Dyson, Development of organometallic (organo-transition metal) pharmaceuticals, *Appl. Organomet. Chem.* **2005**, 19, 1–10.

[14] C.G. Hartinger, P.J. Dyson, Bioorganometallic chemistry – From teaching paradigms to medicinal applications, *Chem. Soc. Rev.* **2009**, 38, 391–401.

[15] G. Jaouen, P.J. Dyson In *Comprehensive Organometallic Chemistry III*; D. O'Hare, Ed.; Elsevier: Amsterdam, **2007**; Vol. 12, p 445–464.

[16] I. Ott, R. Gust, Non platinum metal complexes as anti-cancer drugs, *Arch. Pharm. Chem. Life Sci.* **2007**, 340, 117–126.

[17] P. Köpf-Maier, H. Köpf, E.W. Neuse, Ferrocenium salts – The first antineoplastic iron compounds, *Angew. Chem. Int. Ed. Engl.* **1984**, 23, 456–457.

[18] M. Salmain, N. Metzler-Nolte In *Ferrocenes*; P. Stepnicka, Ed.; John Wiley & Sons: Chichester, **2008**, p 499–639.

[19] K. Strohfeldt, M. Tacke, *Chem. Soc. Rev.* **2008**, 37, 1174–1187.

[20] G. Jaouen, S. Top, A. Vessières, G. Leclercq, M. J. McGlinchey, The first organometallic selective estrogen receptor modulators (SERMs) and their relevance to breast cancer, *Curr. Med. Chem.* **2004**, 11, 2505–2517.

[21] U. Schatzschneider, N. Metzler-Nolte, New principles in medicinal organometallic chemistry, *Angew. Chem. Int. Ed.* **2006**, 45, 1504–1507.

[22] E. Hillard, A. Vessières, L. Thouin, G. Jaouen, C. Amatore, Ferrocene-mediated proton-coupled electron transfer in a series of ferrocifen-type breast-cancer drug candidates, *Angew. Chem. Int. Ed.* **2006**, 45, 285–290.

[23] A. Vessieres, S. Top, P. Pigeon, E. Hillard, L. Boubeker, D. Spera, G. Jaouen, Modification of the estrogenic properties of diphenols by the incorporation of ferrocene. Generation of antiproliferative effects in vitro, *J. Med. Chem.* **2005**, 48, 3937–3940.

[24] P. J. Dyson, Systematic design of a targeted organometallic antitumor drug in pre-clinical development, *Chimia* **2007**, 61, 699–703.

[25] S. J. Dougan, P. J. Sadler, The design of organometallic ruthenium arene anticancer agents, *Chimia* **2007**, 61, 704–715.

[26] A. F. A. Peacock, P. J. Sadler, Medicinal organometallic chemistry: Designing metal arene complexes as anticancer agents, *Chem. Asian J.* **2008**, 3, 1890–1899.

[27] M. Suggitt, M. C. Bibby, 50 years of preclinical anticancer drug screening: Empirical to target-driven approaches, *Clin. Cancer Res.* **2005**, 11, 971–981.

[28] A good starting point for information on the HeLa cell line and Henrietta Lacks is Wikipedia: www.wikipedia.org

[29] G. Bernhardt, H. Reile, H. Birnböck, T. Spruß, H. Schönenberger, Standardized kinetic microassay to quantify differential chemosensitivity on the basis of proliferative activity, *J. Cancer Res. Clin. Oncol.* **1992**, 118, 35.

[30] S. C. Dkhara, Rapid method for the synthesis of cis−$[Pt(NH_3)_2Cl_2]$, *Indian J. Chem.* **1970**, 8, 193–194.

[31] G. B. Kauffman, D. O. Cowan, Cis-and trans-dichlorodiammineplatinum(II), *Inorg. Chem.* **1963**, 7, 239–245.

[32] P. Skehan, R. Storeng, D. Scudiero, A. Monks, J. McMahon, D. Vistica, J. T. Warren, H. Bokesch, S. Kenney, M. R. Boyd, New colorimetric cytotoxicity assay for anticancer-drug screening, *J. Natl. Cancer Inst.* **1990**, 82, 1107–1112.

Glossary

A	adenine
AAS	atomic absorption spectroscopy
acac	acetylacetonate
AMP	adenosine monophosphate
APS	ammonium persulfate
ASS	acetyl salicylic acid
ATP	adenosinetriphosphate
BNPP	bis(nitrophenyl)phosphate
Boc	tert-butoxycarbonyl
BOP	benzotriazol-1-yl-oxyphosphonium
bpy	$2, 2'$-bipyridine
C	cytosine
CAS	chemical abstracts service
CD	circular dichroism (spectroscopy)
cGMP	cyclic guanosine monophosphate
CORM	CO releasing molecule
Cp	cyclopentadienide, $[C_5H_5]^-$
Cp*	pentamethylcyclopentadienide, $[C_5(CH_3)_5]^-$
CT	calf thymus (DNA), also: charge transfer
CT	charge transfer, also: calf thymus (DNA)
CV	cyclic voltammetry (electrochemical method)
CV	crystal violet (an organic dye used to determine cell numbers)
Cys	cysteine
Da	mass units commonly used in biopolymers ($1 \, Da = 1 \, a.u.$)
DCC	dicyclohexylcarbodiimide
DCM	dichloromethane
DEC	1-(N,N-dimethylaminopropyl)-3-ethylcarbodiimide hydrochlorid ($=$ EDAC)
DIPEA	diisopropylethylamine
DMEM	Dulbecco's modified eagle medium (cell culture medium)
DMF	dimethylformamide

DMSO	dimethyl sulfoxide
DNA	desoxyribonucleic acid
dppz	dipyridophenazine
DTT	dithiothreitol
EDAC	1-(N,N-Dimethylaminopropyl)-3-ethylcarbodiimide hydrochlorid (= DEC)
EDTA	ethylenediamine tetra acetic acid (usually used as the disodium salt, so beware of the molecular weight!)
EI	electron ionization (mass spectrometric method)
Enk	Enkephalin
ER	estrogen receptor
ESI	electrospray ionisation (mass spectrometric method)
ET	electron transfer
EtBr	ethidium bromide
FAB	fast atom-bombardment ionisation (mass spectrometric method)
FAD	flavin adenine dinucleotide
FcH	ferrocene
FCS	fetal calf serum
Fmoc	fluorenyl methoxy carbonyl
FDA	Food and Drug Administration
G	guanine
GC/MS	gas chromatography/mass spectrometry
Gly	glycine
GTP	guanosine triphosphate
HATU	2-(1H-7-azabenzotriazole-1-yl)-1,1,3,3-tetramethyluronium hexafluorophosphate methanaminium
HBTU	2-(1H-7-benzotriazole-1-yl)-1,1,3,3-tetramethyluronium hexafluorophosphate methanaminium
HEWL	hen egg white lysozyme
HMBA	hydroxymethylbenzoic acid (a common linker for peptide synthesis resins (cleaved by base))
HO	heme oxygenase
HOBt	1-hydroxy-benzotriazole
HOMO	highest occupied molecular orbital
HPLC	high-performance (or high-pressure) liquid chromatography

IC_{50}	inhibitory concentraion 50%
IR	infrared (spectroscopy)
IUPAC	International Union of Pure and Applied Chemistry
Lc	liquid chromatography
Leu	leucine
LUMO	lowest unoccupied molecular orbital
MALDI	matrix-assisted laser desorption ionisation (mass spectrometric method)
MF	methylenefuranone
MLCT	metal-to-ligand charge transfer
Mmt	methoxytrityl (an acid-labile protecting group)
MO	molecular orbital
MOPS	3-(N-morpholino)propane sulfonic acid
MPE	methidiumpropyl-EDTA
mRNA	messenger RNA (to which the genetic information from the DNA is transcribed)
MS	mass spectrometry
MTP	microtiter plate
MTT	3-(4,5-dimethylthiazol-2-yl)-2,5-diphenyltetrazolium bromide (organic dye used in cell viability assays)
N–N	chelating bidentate nitrogen ligand
NADPH	nicotinamide adenine dinucleotide phosphate (reduced form)
3-NBA	3-nitrobenzylic alcohol
NBT	nitrotetrazolium blue
NMP	N-methylpyrrolidine
NMR	nuclear magnetic resonance (spectroscopy)
NOS	NO synthase
eNOS	endothelial NOS
iNOS	inducible NOS
nNOS	neuronal NOS
PAGE	polyacrylamide gel electrophoresis
PAP	purple acid phosphatase
pBR322	first artificial plasmid (circular double-stranded form of DNA), p stands for plasmid, and BR for Bolivar and Rodriguez, its Mexican creators
PBS	phosphate-buffered saline
PDB	Protein Database
PFA	paraformaldehyde

Phe	phenylalanine
phen	1,10-phenanthroline
PS	polystyrene
pyBOP	benzotriazol-1-yl-oxytripyrrolidinophosphonium hexafluorophosphate
RNA	ribonucleic acid
ROS	reactive oxygen species
salen	N,N'-bis(salicylidene)-1,2-ethanediamine
SDS	sodium dodecylsulfate
sGC	soluble guanylyl cyclase
SNP	sodium nitroprusside
SOD	superoxide dismutase
SPPS	solid phase peptide synthesis
SRB	sulforhodamine B
T	thymine
TAE	tris-acetic acid-EDTA (buffer)
TBAPF$_6$	tetrabutylammonium hexafluorophosphate
TBTU	N,N,N',N'-tetramethyl-O-(benzotriazol-1-yl)uronium tetrafluoroborate
TE	tris-EDTA (buffer)
TEMED	N,N,N',N'-tetramethylethylenediamine
TFA	trifluoro acetic acid
TIS	tri(isopropyl)silane
TLC	thin-layer chromatography
tpm	tris(pyrazolyl)methane
TRIS	tris(hydroxymethyl)aminomethane
tRNA	transfer RNA (which carries the incoming amino acid during protein biosynthesis)
Tyr	tyrosine
UV	ultraviolet
Vis	visible
XO	xanthine oxidase
XTT	sodium 3'-[1-(phenylaminocarbonyl)-3,4-tetrazolium]-bis(4-methoxy-6-nitro)benzene sulfonic acid hydrate (organic dye used in cell viability assays)

Index